JOURNAL OF WILLIAM FOWLER PRITCHARD

William Fowler Pritchard (1818-1875)

JOURNAL OF WILLIAM FOWLER PRITCHARD

INDIANA TO CALIFORNIA 1850,
RETURN VIA NICARAGUA 1852

Edited by
Earl H. Pritchard
and
Phil Pritchard

With an Introduction by
Phil Pritchard

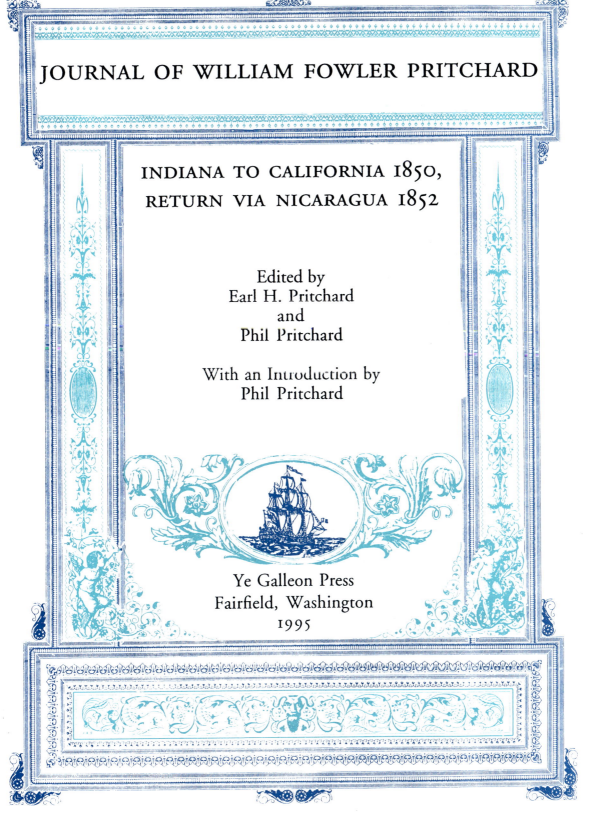

Ye Galleon Press
Fairfield, Washington
1995

Library of Congress Cataloging-in-Publication Data
Pritchard, William Fowler (1818-1875).
Journal of William Fowler Pritchard: Indiana to California 1850,
Return Via Nicaragua 1852; edited by Earl H. Pritchard and Phil Pritchard.
Includes bibliographical references, maps, illustrations, and indices.
1. Pritchard, William Fowler, 1818-1875 – Diaries.
2. Overland journeys to the Pacific.
3. West (U.S.) – Description and Travel – 1848-1860.
4. Pioneers – West (U.S.) – Diaries.
5. California Trail.
6. New Harmony, Indiana – History.
I. Title.
Pritchard, Earl Hampton (1907-).
Pritchard, Philip Norman (1948-).
ISBN ...   0-87770-569-0
ISBN ... pbk. 0-87770-570-4

Ye Galleon Press, Box 287, Fairfield WA 99012

New York, NY
First Edition 1995

Printed in the United States of America
99 98 97 96 95   5 4 3 2 1

CONTENTS

v

ILLUSTRATIONS

PREFACE

My father and I would like to thank, first of all, the many descendants of William Fowler Pritchard who have waited, literally for decades, for this journal to be published. The first, handwritten, transcription of the journal was prepared by my father in the 1970's and then typed. I became involved, primarily to make the revisions we knew would be necessary easier to manage, and entered the text into my first computer, a North Star Horizon running CP/M in about 1982. That version was also the first with notes. My father found it increasingly difficult to make progress on the journal and I eventually took the work over, thinking it would be a simple thing to add a few more notes and get it done. That was in 1988 and the journal has gone through ten versions on my three Apple Macintoshes since then. The "few more" notes, introduction, maps, and indices now comprise nearly half of the total text.

We would also like to thank Mrs. Josephine M. Elliott of the Workingmen's Institute in New Harmony, Indiana, who did research for my father in the Institute's invaluable Local History Card File, for which we must also thank the many librarians and staff of the Institute over the years. Mary Agnes (Pritchard) Zampino provided me with a copy of *The Angel and the Serpent*. John Franklin Pritchard and Edith (Toth) Pritchard found the record for our common ancestor's arrival in New Orleans in 1842. Thanks are also due to the Research Library of the New York Public Library. I would like to thank my wife, Mary Dohnalek, and Bob Gruska for reviewing parts of the final and near final versions and offering many useful critiques and bits of information which were incorporated into the final version.

PHIL PRITCHARD

INTRODUCTION

ILLIAM FOWLER PRITCHARD was born in Ellesmere, in the county of Shropshire, England, on the northern part of the Welsh border, on 24 June 1818.[1] His father, Thomas, was a maltster[2] and his grandfather, Peter, a solicitor.[3] His mother was Juliana Fowler, probably the daughter of a William Fowler from Newcastle-Under-Lyme in neighboring Staffordshire.[4]

William married Florence MacDonagh, daughter of a wine merchant from Chesterfield, Derbyshire, who was born in that town on 5 May 1819. Little else is known of her father's family; a bit more is known about the family of her mother, Dorothy Needham, which had been established in Chesterfield for several generations.[5]

Several years before William and Florence departed for the United States, they moved, with at least William's father Thomas, to Liverpool.[6] There their first two sons were born: Henry Turner Pritchard on 12 November 1838, and Thomas MacDonagh Pritchard on 7 August 1841.

Like many Englishmen of his time, William Fowler Pritchard emigrated to the young United States. Although we have no direct evidence of the reasons for his emigration, his journal of his trip to California and back during the years 1850-52 shows that he was concerned about his growing family's economic future, and this may have been the main reason for his leaving the village in which his ancestors had lived for at least five prior generations and in which he had dozens of relatives. He also makes remarks in the journal suggesting that the freedom enjoyed by the citizens of the United States was attractive to him; he especially remarks on Europe's "tyranny and oppression" in the entry for 4 July. Several of the remarks in his journal clearly indicate that he considered his native England a participant in this tyranny and oppression. In the entry for 20 August 1850 he expresses happiness that a "colonist" had whipped "John Bull" in a fight, and in the entry for 4 September 1852, he says of Jackson's defeat of the British at New Orleans that he "gave the British just what they wanted; a good beating and may they always get the same when they try to trample on the people's liberty."

The Pritchard family arrived in the United States at New Orleans on 22 June 1842 aboard the Governor Davis of Boston from Liverpool.[7] A third son, William Austin Pritchard, was born on 19 November 1843 in New Orleans. Both William Austin and Henry Turner died within two years, after which the family moved up the Mississippi, eventually settling in New Harmony, Indiana in 1847 with a fourth son, William Shakespeare Pritchard (born on 1 December 1845 in Covington, Kentucky).

New Harmony must have been an interesting place at this time. This tiny town had been the site of not one but *two* utopian experiments. It had been founded in 1814 by the Harmony Society, a German millenarist sect (one of the Lutheran Separatist sects) led by George Rapp after they abandoned their first site in America at Harmonie, Pennsylvania. They built a very orderly, celibate, industrious, and religious utopia at New Harmony and prospered by trading their manufactures down the Wabash River on which the town is located. By 1824 they had decided to move back to their original center in Pennsylvania (but at a new site, called Economy, now Ambridge) and sold the town to the English utopian socialist industrialist Robert Owen who established a "Community of Equality" there. He imported a "boatload of knowledge" (a collection of leading scientists, but also literally a boatload, because most of them arrived in a single boat) and planned to build a utopian community for workers on the site. Owen's schemes for a highly structured utopia on the Wabash did not occupy him for long.[8] After Owen returned to England, William Maclure, a noted geologist, became the town's leading figure, focusing more on education and science than on social engineering.

By the time William and Florence arrived, New Harmony was a very small town without any utopian pretensions but with a wholly disproportionate share of intellectuals; the town was the site of the U.S. Geological Survey for 17 years (1839-1856). The language used by William in his journal, although it contains a number of mistakes, is considerably above the average for the times, and he may have been attracted to the town because of its reputation for intellectual and cultural life. He was certainly interested in the theater (recall that he named his fourth son William Shakespeare Pritchard) and found many like-minded people in New Harmony where he became a pillar of the local theater company, founded by William Owen in 1828.[9] William Fowler Pritchard pursued his theatrical interests even in California, where he made his (presumably very modest) fortune working in the theater.

The couple's fifth son, Julian Peter Pritchard, was born 1 March 1848 in New Harmony.

When gold was discovered in vast quantities at Sutter's mill in California, William Fowler Pritchard, like many other Americans, saw a chance to make his fortune and, perhaps, to satisfy his restless spirit and desire for adventure. In spite of the fact that Florence was pregnant with their sixth child (John Fowler Pritchard, born 20 October 1850 in New Harmony while his father was in California), he undertook the perilous journey (which may not have been perceived as quite as perilous as it actually was at this early stage in the Gold Rush when information was sketchy at best) with others from New Harmony. The fact that he kept a journal in which he showed a wide-ranging interest in all sorts of people, places, and events, indicates that he may well have wished to have an adventure, or at least go exploring, one more time before settling down to the safe but relatively boring task of being a father (they were to have four more sons) and cabinetmaker in the little town on the Wabash, by then well within the tamed areas of the United States. In his letter home (included below), he says "I don't think I could like to live in that picayune place," referring to New Harmony, while his remark in the entry for 12 September ("I would not come through again for all the wealth of California") suggests that any urge for adventure he may have had was more than satisfied on the trip.

The journal itself suggests that the family was not terribly well off even though they owned (or were later to own) one of the original houses of the Rappites in New Harmony.[10] William and his wife seem to have had some sort of agreement that he was going to make enough money to get them properly started in his usual business and then return: he complains about missing the opportunity to make much more money in his letter home. It is clear that he had no wagon or horse of his own on the trip (he reports attempting unsuccessfully to buy or trade for a horse or pony on several occasions) and was somehow tied to the original leader of the train, Mr. Sweasey, although he came to have extreme contempt for him.[11]

Other aspects of William Fowler Pritchard's character are not very clearly revealed by what he wrote in his journal. He seems to have had somewhat mixed (but, on balance, negative) feelings about the Native

Americans the train encountered and had almost no cause to even mention African-American travelers.[12] He remarks on 26 June that a man charged with shooting another man for insulting his sister ought to go free while the shot man ought to die, a judgment we might think severe today. He takes a different attitude to another practice concerning honor, the fighting of duels; he calls those participating in such duels "fools" in his entry for 14 August 1852. His remarks about preachers on 15 and 22 August 1852 suggest he was willing to be critical of religion or at least of some of its officials.

After returning from California, he did indeed set up in business as a cabinetmaker (and is known to have built at least one hearse and served as the town's undertaker as well).[13] More sons were born: the twins Volney Walter and Florian Percival Pritchard on 27 May 1853, and their last child, Norman Lockley Pritchard on 4 November 1859. He served in the Regimental Band of the 25th Indiana Volunteers during the Civil War but was sent home early and mistakenly included on a list of deserters for many years. His fourth son William Shakespeare Pritchard served with Company D, 74th Indiana Regiment, which was with Sherman on his famous march to the sea.[14]

William Fowler Pritchard died across the Wabash River in Albion, Illinois at the house of his son William Shakespeare Pritchard on 12 August 1875. Florence was active in the New Harmony Episcopal church and died in New Harmony on 9 August 1896. They are both buried in Maple Hill Cemetery just south of New Harmony. To date, William and Florence have more than 300 known descendants through the six of their nine sons who had children.

James Bennett's Journal

The journey made by William Fowler Pritchard's party is unusual in that more than one description of the trip was published. James Bennett's journal of the trip was published in the *New Harmony Times* between 16 March and 3 August 1906; it was reprinted as a pamphlet entitled *Overland Journey to California: Journal of James Bennett Whose Party Left New Harmony in 1850 and Crossed the Plains and Mountains Until the Golden West Was Reached* in an edition of 200 in New York by Edward Eberstadt[15] and reprinted again by Ye Galleon Press in 1987.

William Fowler Pritchard had a friend named James Bennett in New Harmony with whom he was engaged in theatrical pursuits.

William Fowler Pritchard's remarks in the entry for 6 September 1850 make it almost certain that this was not the James Bennett who published this other account of the trip. The James Bennett with the wagon train is only mentioned three times by William Fowler Pritchard, who, in turn, is mentioned only three times by Bennett in his account, which supports the idea that they were not the best of friends.[16] The two remained with the same portion of the initial party after it split up on 21 July, so their accounts are directly comparable until William Fowler Pritchard began packing on his own over the Sierra Nevadas on 17 September.

The editor of the *New Harmony Times* included a foreword to Bennett's account which is of questionable accuracy at best. Either the editor's list of those who left from New Harmony is far too short or Bennett himself is very lax in listing the large number of people who must have joined in the early stages because half of the party has not been named by the time they crossed the Platte on 16 June when a count of the party is reported (see *The Party* below for more details on this subject). The foreword also has spellings of names which are often at odds with those used in the rest of the work, although it does provide first names which are largely lacking in both Bennett's and William Fowler Pritchard's accounts.

A substantial number of the notes to the text which follows indicate where Bennett's account differs from that of William Fowler Pritchard. The general conclusions to be drawn from these comparisons are as follows:

The spellings of peoples' and places' names is inconsistent between the two accounts: Carlisle/Carlysle, Combe/Combs, Dexter/Decstor, Pullyblank/Pulabank, etc. Since either author would have had few if any chances to see these names in print, this is not surprising.

Bennett has more to say about the people on the trip, or at least he is more careful to name them. He identifies 33 individuals as against William Fowler Pritchard's 20 (not counting four William Fowler Pritchard mentions in his letter home whom we can identify with people Bennett mentions). Only Morrison, a fellow Liverpudlian who joined the party on 7 August 1850, is mentioned by William Fowler Pritchard but not by Bennett. William Fowler Pritchard, on the other hand, is much more informative about matters of natural and local history and usually gives more details about any given event. He specifies 34 types of plants and 38 types of non-domestic animals, and names

74 landmarks (37 rivers, creeks, springs, lakes, etc., 26 mountains, rocks, valleys, passes, etc., 8 forts or towns, and 3 others).

Bennett is less concerned to report each day's mileage; William Fowler Pritchard does it almost every day (the numbers often jibe when both men report them) and he also describes how the figures were arrived at (entry for 31 May).

The Party

According to the foreword to Bennett's journal, the party which left New Harmony on 1 April 1850 consisted of James Bennett, Mr. and Mrs. Samuel Bolton, William Bolton, Miles Edmonds, William Faulkner, George Hamilton, Jonathan Jackson, Ira Lyon, John Mills, John O'Neal, Mitch O'Neal, Mr. Otzman, William Fowler Pritchard, Mr. and Mrs. Sweasey, their son Richard (10 years old) and a daughter.[17] This list is misleading because Mr. and Mrs. Bolton (at least) proceeded on their own by river to St. Joseph, Missouri, where they rejoined the train. We can thus identify at least 18 people in 8 wagons at this point.[18]

There were almost certainly more people than this in the train when it left New Harmony. On the trip to St. Joseph, Bennett mentions several people joining the party: Mr. Spencer (3 April, in Albion, Illinois) and Mr. Fewer, Mr. Moore, and Mr. Wade (20 May, in St. Joseph). But other people are mentioned without explicitly stating that they joined the party at the point they are first mentioned, and it is likely that they had been with the party since New Harmony: Mr. Combs (10 April), Mr. Mitchell (15 April), and Mr. Pullyblank (10 May).

After leaving St. Joseph, the train was joined on 24 May by another "company" led by a Mr. Dexter and containing at least Dexter and his wife, a Mr. Corbin[19] (mentioned by Bennett) and one or more Wilseys (mentioned by William Fowler Pritchard – probably Bill Wilsey and his wife). The combined train is reported as having 14 wagons on 28 May. Again, other names crop up without any information about how long they have been with the party and no indication whether they came from New Harmony or were part of Mr. Dexter's party: Mr. Beal (5 June), Mr. Axton (26 June), and Mr. Williams (8 July). We also find out during this period that there is a *Miss* Combs (30 June), presumably a sister or non-minor daughter of the Mr.

Combs mentioned earlier. On 9 June, we are told that George Hamilton has left for another party.

On 16 June, there were 14 wagons and 63 souls crossing the Platte. Since Bennett and William Fowler Pritchard between them identify only 34 people, nearly half the party remain unknown.

The party was decreased by two deaths in July: William Faulkner by drowning on 9 July and his aunt, Mrs. Samuel Bolton, of illness, on 21 July.

On 21 July, the party split. Dexter led the dissidents, but only some of the people already mentioned can be put in one party or the other. A Mr. Hinkley, not mentioned before, was with Dexter. The following are mentioned subsequently by Bennett or William Fowler Pritchard as if they were still with Sweasey's party: Bennett, Fewer, Jackson, Lyon, Moore, Mills, William Fowler Pritchard, Spencer, and Sweasey and family. This party was reduced to 6 wagons. Mentioned as having gone with Dexter are Beal, Samuel Bolton (presumably with his son William), Hinkley, M. O'Neal, Otzman, Pullyblank, Williams, and the Wilseys.

On 7 August, Sweasey's party gained a Mr. Morrison and wife, but only for a while because William Fowler Pritchard reports him as having misfortunes elsewhere on 31 August.

On 24 August, Dexter's party split. On 27 August, Bennett reports that Mr. Pullyblank had also formed his own party – this was probably the group which split off from Dexter's party. Jonathan Jackson of Sweasey's party had strong ties to some of Pullyblank's party and joined 5 or 6 of them to begin packing (i.e. proceeding on foot with just a pack) on 1 September.

On 2 September, a wagon was abandoned, reducing the party to five wagons.

On 10 September, Spencer began packing.

On 15 September, another wagon was abandoned, leaving four. Three of the company are mentioned as having left with "traders," but this may be confused with the later departure of Bennett, Mills, and Lyon.

On 17 September, William Fowler Pritchard left and began packing.

On 18 September, Bennett himself, along with Mills and Lyon, left Sweasey's party to drive a herd for a "speculator in cattle." This left

Sweasey's party with only (so far as we can tell) Fewer, Moore, and Sweasey and family.

William Fowler Pritchard arrived in Sacramento on 27 September and Bennett on 1 October, but we do not know when Sweasey's, Dexter's, or Pullyblank's parties arrived.

In addition to these people mentioned on the trip by William Fowler Pritchard and/or Bennett, there are a number of New Harmony people mentioned in William Fowler Pritchard's letter home who were probably on this trip. Several of these are also mentioned on the "California Argonauts 1850" card from the Local History Card File at the New Harmony Workingmen's Institute: Albert Fisher, Adam and George Lichtenberger, Bob Robson, and James Madison Stoker. Others in the letter can be identified from other cards: O. D. Chaffee, Dr. John T. Cook and Ann Leonora Travers Cook, William Daniel (questionable), Alexander Doyle, Oscar Felch, John Gullett, John Hale, Dr. Daniel Neal (highly questionable), J. O'Neal, Henry Pratton, and James Madison or Richard Stoker (the card is ambiguously worded). Those mentioned in the letter but not on any card include the son and daughter of Combs, Dr. Conyngton and wife, F. Duckwork (J. O'Neal's uncle), W. Evans, and Mr. Wilkinson. This makes 23 further names of possible members of the train (counting only one of the Stokers – see the notes to the letter).

Finally, the California Argonauts 1850 card mentions the following persons for whom we find no references in the journals of the trip: Thomas Cox and wife, John Craddock, his wife Betsy, and a child lost on the trip, Michael Craddock, his wife, and son, Absalom Driggers, Sam Endicott, Richard Ford, Frank Marsh, Dan Perkey, Charles [Augustus] Twigg, J. H. Verriel and family, John William's wife, and a David, a John Jr., and an Ellen who are either Williams or Corbins (this card too is ambiguously worded). This makes 21 more names, but the death of Craddock's child on the trip makes it unlikely that he was really an 1850 traveler with this train since neither account mentions this event which surely would not have gone unnoticed or unreported.[20]

Counting all of these names, we have 78, while the maximum census of the train is given as 63; even discounting the Craddocks, we are 10 over this number. Resolution of the exact makeup of the party must await further researches.

The Trip

There were about 45,000 travelers on the California trail in 1850, nearly twice as many as there had been in 1849, the first year of the "Gold Rush." Only about 1 to 2 percent of these were women and children, making the 4 women and at least 2 children with William Fowler Pritchard's train of 63 an exceptionally high percentage. The events of the journey suggest that there were good reasons for keeping these groups from the journey – both of the fatalities to members of this party fell upon these groups or those close in age to these groups (the "young man" William Faulkner and Mrs. Samuel Bolton).

The party took what was the main route for 1850 (which diverged in three places from the 1849 route by taking Sublette's cutoff, Hudspeth's cutoff, and the Carson River route). The remarks in the entry for 15 August 1850 about rejoining the regular road at the end of Hudspeth's cutoff and again having lots of company might lead one to suspect that they took an unusual route, but this was not the case.

The route was more crowded, overgrazed, and understocked than in 1849 due to the increased traffic, and William Fowler Pritchard's many remarks about having trouble finding grazing for the cattle support this. On the other hand, there were more ferries than there had been a year earlier, and even some bridges.[21] Stewart states that three-quarters of all wagons were abandoned before reaching California, and the large number of abandoned wagons noted by William Fowler Pritchard make this statement credible.

It is interesting to note that although the party met the famous Kit Carson on the trip, William Fowler Pritchard did not think this worthy of mention at the time and only mentioned it on 13 September 1850 when he was discoursing on the naming of the Carson River in Nevada, near Lake Tahoe.

Their train completed the trip in less time than the average; they started late and arrived early (see the note for 12 September 1850).

Returning via Central America, as William Fowler Pritchard did, was a common return trip after 1850 when the route was opened up.[22]

The Manuscript

The manuscript itself consists of six small notebooks and a "Pocket Letter Book." Virtually all of the entries were made in pencil.

Books one and six are 6¾" tall by 4¼" wide, are bound in stiff red cardboard covers with a shallow embossed floral pattern, and contain 22 and 24 medium weight light blue pages respectively, not counting the inside front and back covers. Book one originally also had 24 pages because there is a cut out leaf at its front. Book one's cover is intact, while book six's has been separated. The text of the journal begins on page 5 of book one – there is a bit of text on pages 1-2 from 30 September 1850 (the last words relating to the California-bound part of the trip) and pages 3-4 are blank. Pages 5-7 contain the start of the journal (1 April and 19-23 May 1850) while the rest of the book is either blank or contains "Pritchard and O'Neil Account Book" entries dated 17 June 1852-7 July 1852. Most of the entries are just names and amounts, mostly between 25¢ and $1.25, with about seven entries per day. It would be most consistent with the dates of these various sections to find that the account book entries (being added later) start at the back of the book and are upside-down relative to the material which begins the book, but there is good reason believe the contrary. The pages are lined and have a light red double rule near the top, one vertical light red rule at the left, and one double and one single light red rule at the right going down from the top horizontal double rule (i.e. the pages are in a single entry account book format). Since the account book entries are right side up with respect to this format, it is most likely that they, not the journal entries, are at the start of the book.

Given these facts, and the fact that book two begins with repeat entries for 20-23 May, it is likely that the actual journal William Fowler Pritchard kept on his trip to California began with book two and that the material in book one was added at some later point, probably after the end of its use as an account book – i.e. after 7 July 1852, just a month before he started on his trip home, perhaps even at a much later date. It is even possible that William Fowler Pritchard compared notes with James Bennett or saw his original journals before making these entries.

Books two and three are 5¾" tall by 3⅞" wide, are bound in thin tan leather, and contain 44 and 38 medium weight yellowed white pages respectively, not counting the inside front and back covers. Book three originally also had 44 pages because there are 3 cut out leaves at its front. Book two ends with some out-of-sequence remarks relating to Brady's Island, seen on 11 June, and Scott's Bluffs (22 June), plus duplicate statistics reported at Fort Laramie (25 June) and a duplicate

after a long preparation we were at
length ready to start & on the
1st of april 1850 rather a bad day
as we did not know but what
we were (april fools or not) we
took leave of our families the scene
I shall not attempt to describe suffice
it to say there was plenty of tears
I prayer for our safety & we crossed
the Wabash & had all over by
night, at night the band (we had
several others with us) played several
airs amongst the rest "Good bye" the
young folks came across & a many
people they danced & then returned
in the morning we were all ready
to start when a lot of the inhabitants
inhabitants came down to the
river & gave us 3 hearty cheers

Book One, Page 5

entry for 27 June. Book three ends with 3 pages upside-down with respect to the rest of the book. These last pages seem to consist almost entirely of lists of costumes and stage props for actors, which are explained by William Fowler Pritchard's activities with the theater in Sacramento.

Books four and five are 6¼" tall by 3⅞" wide, have no covers, and contain 32 (book four) or 64 (book five) medium weight yellowed white pages. Book four is a single signature with a single crudely done stitch in the middle third and may have been made from large sheets by William Fowler Pritchard. In spite of the fact that it begins in mid-sentence, it is an exact continuation of the text from book three.

Book five takes up the description where book four leaves off and finishes the trip to California, except for the last few lines already mentioned which are found in book one.

William Fowler Pritchard's letter home is in a separate booklet, a "Gregory's Express Pocket Letter Book" dated 1851. It is 5" tall by 3⅛" wide, has a thin black glossy cover, and contains 26 thin blue-tinged white pages, guaranteed, with envelope, not to exceed the weight of a single letter. It is stitched with two stitches of about ¾" which overlap one another a bit at the center of the page. These go through the pages, but inset from the edge of the fold and not at the folds as in the other books. Pages 2-3 are blank; the text of page 1 continues (if somewhat awkwardly) on page 4. Pages 14-15 are also blank but the text continues sensibly on page 16. The last leaf is blank and there is an 1851 calendar on the inside back cover. "Gregory's United States and California Express" was associated with Thompson & Hitchcock of 149 Pearl Street (corner of Wall), New York City, although just how is not clear from the advertising on the front and back covers.

Book six is odd. The cover is broken and the leaves have been numbered but possibly not by William Fowler Pritchard (the page numbers seem newer and sharper than the rest of the writing and none of the other books have leaf numbers). The first signature has 24 pages like book one, but leaf 13, the first numbered leaf following this signature, has a hard backing similar to the cover material – but this does *not* appear to be the back cover of the book. If it is, it is very faded and the tear marks do not match those of the front cover. It has the words "Going home via Central America" on its back (i.e. faded red) side. Another signature of 24 pages follows this and is of the same type of paper. The journal's text ends on the front side of leaf number 19

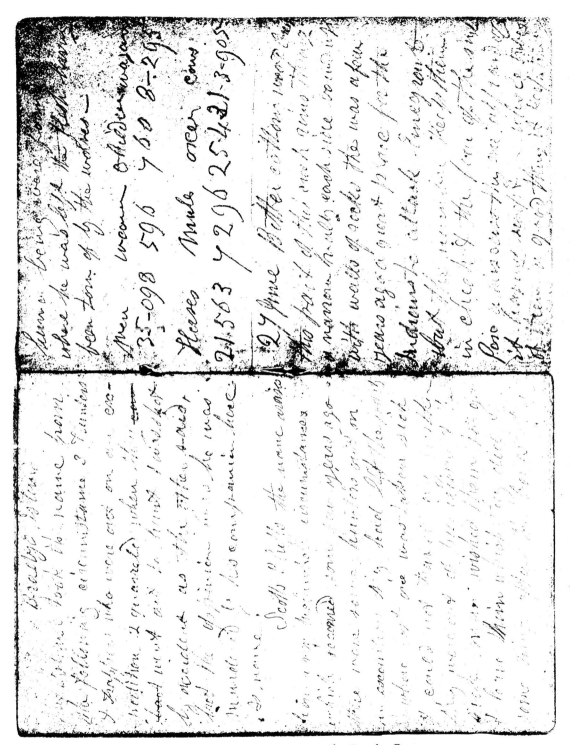

Book Two, Page 44 & Inside Back Cover

(page 38) and is followed by a variety of other material. This begins with some notes on plants wanted for a garden, a list of seeds "at home," two lists of some lumber, one for a fence, the other for a person whose name is illegible, a list of some expenses for Julian Pritchard (one of William Fowler Pritchard's sons) dated 7 December 1872, more lists of garden seeds, dated 1873 and 1878 (after William

GREGORY'S CALIFORNIA EXPRESS.

THIS LINE, one of the oldest established in the business, d[is]patch messengers TWICE EVERY MONTH, by Steamers leavi[ng] New-York and San Francisco, in charge of Letters, Parce[ls] Packages, Gold Dust and valuables, for distribution through out the United States and California. BEING INVARIAB[LY] AHEAD OF THE MAILS, the best medium is presented to [the] public for the prompt dispatch of correspondence.

If possible, letters should be directed to the care of mercant[ile] houses, or well-known residents, in San Francisco or o[ther] parts of California, by which means their earlier delivery m[ay] be insured.

Letters directed simply to "California," or "at the mine[s]" will not be forwarded, as little probability exists of their rea[ch]ing the persons so vaguely addressed.

After addressing the letter, enclose it in an envelope, a[nd] direct to J. W. GREGORY, 149 Pearl Street, New-Yo[rk,] post-paid to New-York. Messengers leave this office, by eve[ry] Steamer, for San Francisco, with letters, small parcels, a[nd] packages, to be delivered immediately on arrival at San Fra[n]cisco, to MR. JOSEPH W. GREGORY, who has every facili[ty] at hand for their immediate delivery to the consignees in S[an] Francisco and various parts of California. The arrangeme[nt] of this Express being of the most perfect description, no del[ay] can occur that it is possible to provide against. Parcels receiv[ed] until the morning of the day the Steamer leaves, and lette[rs] until 2½ P. M.; packages and heavy freight must be delive[red] at our Office prior to the day of sailing, where goods awaiti[ng] shipment are covered by insurance.

THOMPSON & HITCHCOCK,

MANAGERS & AGENTS,

149 Pearl Street, cor. W[all]

P. S. *Shippers* by this Express, can avoid the annoyance a[nd] expense of clearing their goods at the Custom House. Ter[ms] reasonable.

Tickets for the different Steamers can be procured at t[he] Offices of this Express in New-York, San Francisco, or Sa[c]ramento City,

Information concerning the forwarding of goods, or the d[e]parture of the different lines of Steamers, furnished to appl[i]cants, by mail or otherwise.

My very dear wife,

I received your letter dated the 28th April, post mark 1st May. it stated that you had recei ved a draft. but for what amount it said nothing at all about it. this is very wrong it makes me feel un comfortable. you speak as if you had

Letter Home

Fowler Pritchard's death!), 2 blank pages, a note about materials received from another person (whose name is also illegible) dated 1872, records of hours worked by two of his other sons (Florian and Volney) and a William Warren, and one page of indecipherable accounts. Following this is the back cover whose tear marks and creases match those of the front cover and whose inside surface has writing (upside-down with respect to the rest of the book) containing lists of clothes, food, and other possessions. It appears that the original book six and another of the same type were deliberately broken and that the pages from the other book (one of its covers having been thrown away or subsequently lost) were inserted following a page consisting of one of the covers of the second book whose inside surface had been used to continue the text, and that the original book's back cover was then appended to the result. The inserted page has partially missing corners whose appearance suggest it may be a paste-up of a regular page and cover material, but this is uncertain.

With the journal proper there was also a booklet, 5" high by 3" wide, with no cover, consisting of 40 thin faintly horizontally-lined white pages, sewn in the middle through the fold and with a straight pin to hold it together as well, with gold leaf on all three edges. It contains a variety of miscellaneous material, probably in several hands and from different periods, including some accounts from 1866 and relating at least in part to William Fowler Pritchard's and William Shakespeare Pritchard's service in the Union Army during the Civil War.

Editing Rules

The manuscript is very badly punctuated and capitalized, and changes to bring the text up to a normal standard in such matters constitute the most common changes from the original. Punctuation is quite sparse, and almost all changes consist of additions. Most sentences did not begin with a capital letter and many proper names were not capitalized. On the other hand, in accordance with the style of the 19th Century, many words which we would not capitalize today were capitalized, and these have been made lower case.

In some cases, verbs were changed from singular to plural where this would have been correct usage: "there was many Indians" was changed to "there were many Indians" and so forth. In a number of instances William Fowler Pritchard uses an adjective where an adverb is appropriate and these have been changed ("tolerable good" was changed to "tolerably good," etc.) Other minor corrections (which are not always necessary) such as substituting "neither... nor" for his "neither... or," have also been made.

Archaic or phonetic spellings have been changed to the modern accepted ones to make the places involved clearer to the modern reader; where this has been done, the actual text is shown in a note. In addition, British usages such as "colour" have been changed to American usages such as "color;" William Fowler Pritchard was not consistent in his use of such Britishisms. Many words which are today single words appear in the manuscript as two words ("water falls," "some one," "day light," "raw hide," etc.) and these have been converted to modern usages. There are a considerable number of simple spelling errors ("birth" where "berth" is clearly meant, "teem" where "team" is correct, "slew" for "slough," "ate" for "eat," "off" for "of," etc.) and these too have been corrected.

In a few cases, a word has been interpolated for the sake of readability or clarity; all such additions are enclosed in square brackets. The entire narrative from 2 April to 19 May 1850 is a summary from Bennett and is similarly enclosed in brackets. The indicators of the starting place of each book are also bracketed since they are not a part of the actual text in the journals.

In spite of all these changes, the intent has been to preserve the distinctively non-modern style of the text, only smoothing out the rough edges whose retention would not contribute to this aim.

Illinois

● 1 April
New Harmony

[Book One of *Journal*]

fter a long preparation we were at length ready to start and on 1 April 1850[23] (rather a bad day as we did not know but what we were all April fools or not) we took leave of our families. The scene I shall not attempt to describe; suffice it to say there were floods of tears and prayers for our safety, etc.[24] We crossed the Wabash and had all over by night.[25]

t night the band (we had several of them with us) played several airs, amongst the rest "Good-bye." The young folks came across and a merry people they danced and then returned. In the morning we were all ready to start when a lot of the inhabitants came down to the river and gave us three hearty cheers. We returned them and then turned our backs on New Harmony, [Indiana][26] our homes, families, and friends, perhaps forever. The last object we saw was the old saw mill and at the turn of a road we gave it a parting cheer, turned the corner, and all was out of sight.

[After the preceding account of the first day, the Pritchard *Journal* jumps to 19 May in the vicinity of St. Joseph, Missouri. The main events of the seven weeks journey across Illinois and Missouri are briefly summarized from *Overland Journey to California: Journal of James Bennett* [27] and are contained in brackets as are all departures from the text of the Pritchard manuscript.

In addition to the people who are mentioned as being on the trip by William Fowler Pritchard, the Foreword to Bennett's journal states that William Bolton, Miles Edmonds, Ira Lyon, John O'Neal, and Mrs. W. J. Sweasey were with the party when it left New Harmony. These people are all mentioned in William Fowler Pritchard's letter home with the exception of Miles Edmonds,[28] and details of their lives will be found where they are mentioned.

For two days they travelled north through Fox River, French Creek bridge, Grayville and Albion, Illinois[29] (reached 3 April, about eighteen miles), where a Mr. Spencer joined the party.[30] After this their direction shifted to west-northwest via Maysville (6 April), Twelve Mile Prairie (7 April), and Grand Prairie (9 April) to Carlyle[31] (forty-five miles due east of St. Louis, 10 April), where they crossed the Okaw River.[32] Their leader was W. J. Sweasey.[33] Bennett as chief scout and hunter travelled ahead, generally accompanied by either Lyon,[34] Mills,[35] or Combs.[36] They hunted for small game (squirrels, rabbits, geese, quail, etc.) and deer (one killed on the seventh and another on the tenth) and kept the company pretty well supplied to St. Joseph. The weather was unusually cold (indeed, it snowed on 9 and 14 April and then again on 7 May) and frequent rains often turned the poor dirt roads into quagmires, limiting the ox teams to ten to seventeen miles a day.[37] Although telegraph wires were seen and towns encountered every few miles, the party felt it necessary to keep watch regularly.

A few miles beyond Carlyle the St. Louis road forked and the wagons turned west-northwest toward Alton (nineteen miles north of St. Louis, on the Mississippi River), which was reached on 13 April after passing through Looking Glass Prairie (four miles east of Troy, 11 April), Troy (nineteen miles east-northeast of St. Louis, 12 April), and Edwardsville (nineteen miles northeast of St. Louis). At this point they changed plans and headed 100 miles further to the north due to a lack of feed for their animals across the Mississippi at Alton. A strong west wind and cold frosty weather made travel miserable as they proceeded on to Jerseyville (nineteen miles north-northwest of Alton) in a snow storm on the fourteenth. In the morning (15 April) Hamilton,[38] Lyon, and Bennett played some music in a tavern and after the weather cleared the caravan went on past Kane[39] (6 miles north-northwest of Jerseyville) to camp in a wood near a creek. Mr. Mitchell's[40] wagon cover caught fire, but was extinguished without great damage. They remained in camp the next day because of rain but on 17 April they travelled on north through Carrollton (eight miles) and then turned west to encamp five miles from the Illinois River. Carrollton was a flourishing village with two printing establishments, a court house, numerous mercantile shops, and a good brass band. Much of the eighteenth was consumed in getting to and across the Illinois River and its tributaries and finding the proper route northwestward toward Atlas (twenty-four miles north-northwest of where they crossed the Illinois River), the approach to the Mississippi River crossing at Louisiana, Missouri (twenty-five miles down the Mississippi from Hannibal). Thirteen squirrels and a goose were their game for the day, but on the nineteenth Mills killed a deer and they reached Bay Creek.[41] On the twentieth they reached Atlas. Local people were attracted to their camp by singing, dancing, and recitations. On 21 April they crossed the Mississippi on a steam ferry. It took three hours to cross the whole train of eight wagons, ox teams, horses and other cattle.

T he weather turned warm and sultry and a rain storm drenched them just after they had started their evening camp fires about six miles west of Louisiana. From Louisiana their route was mainly west but slightly north toward St. Joseph, Missouri. It passed through two small villages, Spencerburg[42] (seventeen miles due west of Louisiana) and Madisonville[43] (Tuesday, 23 April), crossed the Salt River[44] and reached Paris (thirty-two miles east-northeast of Spencerburg) on 25 April, where the weather turned warm again after several very cold days, and thence on through Milton[45] (where some wagons were repaired) and Huntsville (twenty-eight miles west of Paris) on 27 April. Rain and mud delayed them so that on the evening of the twenty-eighth, when they were much intrigued by the native song of a party of Germans, they were only a few miles beyond Huntsville. They were dissuaded from leaving the main road and going south to Glasgow and were told that 765 wagons had already passed that way that year. Two more days of travel brought them to Keytesville[46] (twenty-six miles west of Huntsville) on 30 April.[47] The evening brought them to Indian Grove,[48] and some eighteen miles later, on 2 May, they reached the Grand River[49] ferry which took three hours to cross the 200 yard wide river. The roads in this area were among the worst they had encoun-

5

tered and game was in short supply. The third of May was rainy and they made fourteen miles and then only thirteen on the fourth, when they found good grass for the cattle at Shoal Creek[50] and so decided to spend 5 May in camp to let the cattle graze. In the evening their musicians attracted visitors, and their German friends treated them to folk songs again. On 6 May they covered fifteen miles and ran out of flour but on the seventh they stayed in camp because of snow which lasted for several hours. They reached Kingston (forty-four miles northeast of Independence and what would later be Kansas City) on 8 May after a long drive of seventeen or eighteen miles. One of their spring wagons[51] upset in a mud hole with the ladies in it, but no one was seriously injured. Since their cattle needed rest and feed (which was reported in short supply in St. Joseph), on 9 May they drove about four miles straight north to a good grazing area on Shoal Creek where it was decided to stay for a week.

As some doubted the advisability of the delay and others were disquieted about reported conditions at St. Joseph, where the journey into the uncivilized frontier began, Mr. Sweasey, the leader of the expedition, told the company the next day, 10 May, that he intended to ride on to St. Joseph and scout out the situation. He and Mr. Pullyblank,[52] another leader of the expedition, promptly departed on horseback and this action seemed to quiet the dissatisfaction. On 14 May the two men returned with good reports, and preparations were made for instant departure. They made fourteen miles on the fifteenth but only ten on the sixteenth because they stopped in Plattsburg (twenty-two miles west-southwest of Kingston) to lay in their supply of flour for the journey west.[53] On the seventeenth they made about four miles because much time was spent in making a new wagon wheel. On the eighteenth they crossed the hundred yard wide Platte River[54] on a solid rock bottom where the water reached only to the knees of the cattle and encamped four miles from St. Joseph. On the morning of 19 May three of their best oxen were missing and the whole day was spent in finding them so that they drove into St. Joseph only on Monday, 20 May.[55] Here they were joined by Mr. Samuel Bolton[56] and his wife, who had come by prearrangement by river boat from New Harmony, arriving three weeks earlier. Three other men, Moore,[57] Wade,[58] and Fewer[59], applied to them for passage to California and were accepted.]

After seven weeks of bad roads and wet, snowy, muddy, frosty weather we arrived at Saint Joseph, Missouri, the grand starting point, on 19 May.[60] We then took in our provisions and got everything put in repair. Had three cows stolen – and on 20 May 1850 we crossed the Missouri river in a steam ferryboat. We had to wait all day for our turn, there were so many crossing.

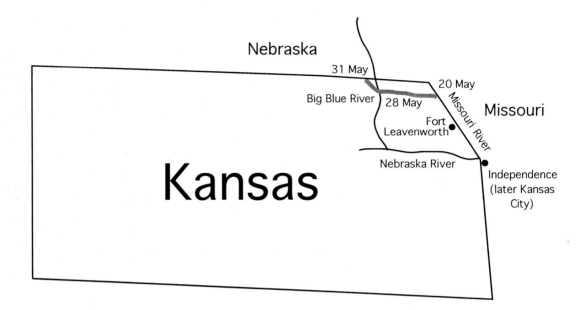

Nebraska

31 May

20 May

Big Blue River   28 May

Missouri River   Missouri

Fort Leavenworth

Kansas

Nebraska River

Independence (later Kansas City)

 e drove on six miles until we came to what is called the bluffs,[61] a range of rocks where the river reaches at high water. We then camped for the night. We are now out of civilization, being in the Indian territory. This part belongs to the Kickapoo tribe.[62] This is a great place for rattlesnakes.[63] One man killed twenty-four. I stood guard.[64]

21 May – Saw a few Indians. A pretty country, nothing worthy of note today. Drove about six miles and then camped. Grass scarce, just beginning to shoot.[65]

22 May – Same as yesterday. Drove sixteen miles.[66]

23 May – Heard that the cholera was ahead. Saw several graves of emigrants. Went eighteen miles.[67]

[Book Two of *Journal*]

24 May – Went twenty miles. Several dead cattle on the road; also several graves, one with all the clothes of the dead person lying near the grave and another covered over with elk horns. A last year's grave.[68]

25 May – Made about twenty-five miles. Caught up with Dexter's[69] party, also two Wilseys.[70] I was very much tired. Had to get up and

ride. They were encamped on a creek called Nemaha.[71] They shot ten or twelve large fish of the buffalo[72] and pike[73] kind.

26 May – Made an early start. Passed a place called the narrows.[74] In descending a hill one of the wagons broke a wheel. We had two wheelwrights and a blacksmith along so we soon had it repaired. We made about fifteen miles and encamped on a sulfur spring. Very hot day and water very scarce so we commenced to carry it along in our canteens over our shoulders.

27 May – Started at sunrise. Made about twelve miles.[75] Should have gone further but there was neither wood nor water. We formed a "corral"[76] for the first time. That is, by putting the wagons in a circle and driving the cattle inside they are easy to catch; also a good place of safety in case of an attack from the Indians. We stand guard three men at a time, four hours.

28 May – A great change in the weather: very cold and windy. Greatcoats that had been put away [are] now in demand. We came up to the grave of an Odd Fellow who, had he died at home, would have been buried in great pomp. As it was he even had no coffin. A piece of board told us who and what he was. His bedding and clothes lay beside the grave. Met several returning on account of sickness; some homesick, some afraid. We came to a river called Big Blue.[77] The stream being very high we could not ford, so we had to make a raft and float the wagons over. We also cut down a cottonwood[78] tree and made a canoe.[79]

29 May – We finished the canoe. Crossed over the women and provisions. The wagon that had my box in [it] had the misfortune to slip off the raft and fall in the river.[80] The raft was made by a company from Virginia. We lent them ropes and helped them and they did the same in return. All my things got very wet and I lost some of my clothes; saved the wagon. We then swam the cattle across and encamped about a mile off on a prairie. Met a company of trappers and hunters from the headwaters of the Missouri and Mississippi and the Rocky Mountains. They had eight wagons loaded with furs, buffalo robes, etc. They were very dirty and ragged, having been in the wilderness some three years.

30 May – Made an early start and made about twenty miles. Saw more graves. No wood or water. Came to the Independence road.[81] The grass [is] scarce; so many [had] passed before. Went about seven

miles farther and encamped for the night. Three people returning camped by us by whom we sent some letters. A cool night.

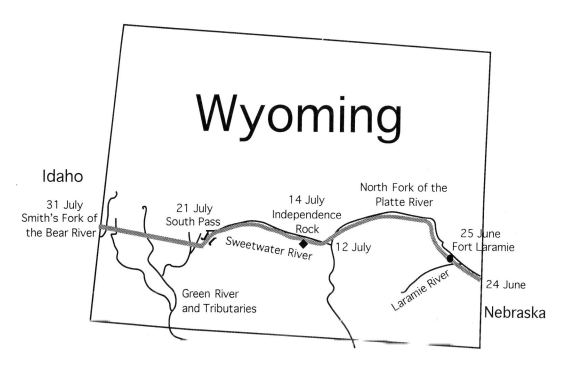

 June – Midsummer Day.[163] This day I am thirty-two years of age.[164] They wanted me to stand treat; as it was raining at the time I told them to help themselves as there was plenty of liquor. We passed a company who had encamped on purpose to bury their dead and attend to their sick. We are within five miles of Fort Laramie. Was on watch last night; moonlight. We have driven our cattle onto an island to feed. Wrote a letter today to my dear wife.[165]

25 June – Shortly after breakfast this morning an old gentleman came to our camp to get assistance to bury a dead man. All their men were sick but one and he was after lost cattle. Myself and four [others] went and dug a grave and carried him to it and covered him over with some willow brush and an old quilt and covered him up. He was from Iowa. The old gentleman was his uncle. He was the second nephew he had buried on the journey. He stood it like a philosopher and dismissed us after repeating the Lord's Prayer. A few miles off, one party lost nine out of sixteen and the sickness here has been dreadful. I never want to go through this dreadful Platte bottom again. A few miles brought us to the Laramie Fork of [the] Platte River, which we had to ford.[166] We had to repeat the same as [at the] Platte River. There

were some men who had tied a rope across the stream and made a boat out of their wagon bed. Four of us paid them $1.25[167] to take our boxes over – they made from $50 to $100 per day. [We are] now in the Crow nation.[168] The stream is narrow but runs very rapidly and what is very remarkable [is that] it rises and falls every day. It comes from the snow in the Rocky Mountains and the sun in the daytime thaws it so in an evening it is high here. After reloading we went on to Fort Laramie, distant about one and a half miles.[169] The fort was formerly a trading post to trade with the Indians and belonged to the St. Louis Fur Company, but when the dispute between England and the United States regarding Oregon was settled and the people wanted to emigrate there, there was a place of safety wanted, also to keep the Indians in check, so this government purchased it and it is now a military station. The fort is built of unburnt brick but there are many more buildings there now. This I expect is the last place we shall see until we arrive in California. We sent off our letters. There is a store here but everything is dear on account of their having to be drawn from the States; dried apples ten dollars per bushel, sugar sixty cents per pound, coffee thirty cents per pound, whiskey $46 per gallon, and so on. The number of persons that have passed here up to date is as follows – men 35,098, women 596, children 760, wagons 8,295, horses 21,563, mules 7,296, oxen 25,431, cows 3,905, number of deaths reported 155.[170] Now we know that the last item is incorrect because one of our party kept an account of all he saw and it is many more and it is impossible for him to have seen all there are. [There are] also a great many wagons and people that never get reported for two companies we know were ahead of us though not reported, and there is also a great many behind sick, and others that have buried some who have not arrived. There is a blacksmith's shop here so some of our wagon wheels having become loose with the heat and dry sandy roads, we left them to be fixed and came along in the morning. There were a great many wagons left there. They were no price at all; indeed they could be picked up all along the road so when we want fire wood at times we burn part of a wagon. They are no use here so far away from anywhere. We see some every day, and broken wheels; also dead cattle. The wolves have fine times. We travelled about twelve miles and encamped on the north fork of [the] Platte again.[171]

26 June – As some of our party started to the fort after the wagons and [as] I had the toothache badly, I went along to get it drawn, but

after getting it lanced the doctor's instruments would not fit, it being a large tooth, so I came away in greater pain than I went. Put in some extract of tar and got relief.[172] Brought back the wagons.[173] Saw at the fort an emigrant in irons for shooting a man for insulting his sister. The man is not dead but ought to be. The other ought to be released. We did not move today but did our washing and let our cattle rest. We heard when at the fort that a wagon upset in crossing at the ford with a woman and child and all their things in [it]. They were saved with great trouble. A company of forty wagons passed our camp; sixty-five men, two Negro women to cook and four yoke of cattle to each wagon and a large quantity of loose cattle.[174] From Missouri.

27 June – After travelling ten miles we came to a most beautiful large clear spring of water but unfortunately it was warm.[175] The cattle would not drink but we drank some as it was a very warm day. The spring is cooler than it was five years ago. After travelling eight miles further we came to Bitter Cottonwood Creek.[176] It was dry but [we] found a spring. Saw a thunder shower coming and tried to make a camp before it came up but it caught us and a pretty drenching we got. The creek was full directly. Made eighteen miles. One of the cows had a calf so it was to be knocked on the head but we fixed up a sack behind one of the wagons and made holes for the legs to come through and it gets along first rate. We did one so before as we came through the States.[177]

28 June – Left Cottonwood Creek and proceeded on our route. We had a tolerable road in the morning and passed through a country of pine timber. Took our dinner beneath the branches of a large pine. It was the first tree we had sat under for some time. The roads were dreadfully bad, rocky, had large holes, and [were] hilly. Had to keep looking at the wheels continually. Saw Laramie Peak;[178] this is the highest of the bluffs on the Laramie Fork. I have seen mountains before but never anything to equal this. As the sun shone on them it was a most beautiful sight to see; peak above peak intersected with rocks, and cedar and pines of a stunted growth here and there add beauty to the scene. We passed Willow Creek.[179] This part of the country abounds with springs of excellent water. Very warm; thunder and lightning, wind, and consequently dust. We passed Horseshoe Creek.[180] This country abounds with elk,[181] deer,[182] buffalo, etc., but the number of emigrants being so great they are driven off the road and it is a toil to go after them. Made twenty miles and encamped in a large

valley where there is plenty of wood and water but little grass for the cattle.[183] Several of them got lame with the bad roads.[184]

29 June – Horrible bad roads up and down all the time; full of rocks. Passed Deer Creek[185] and several smaller ones and tried to get to another creek about six miles further but several of the oxen gave out and dropped down in the team so we had to camp in a dreadfully barren rocky place where there was scarcely a blade of grass or wood or water. We had a little water in our canteens, a very little. The rocks are covered over with wild sage now. If we had the ducks and onions we could do first rate. The feet of the oxen are worn through, so they are lame. A very cold night with wind. Made twenty-two miles.

30 June – Very cold morning with wind and rain. Two of the party went in search of water on horseback with India rubber bags and were successful. A party came by in search of lost cattle. One man had been offered $100 for a horse and the next night the wolves ate it up. Two wolves [were] killed in an adjoining camp. Came to a good spring and cooked dinner. Some of the party went on in search of grass. After dinner we went after the rest of the party and found them encamped on La Prele Creek, a small clear stream.[186] Today the country has been sandy and barren. The road lay, that is a part of it, right in the creek – toads with horns and tails,[187] also a kind of grasshopper of a dirty red color in great numbers.[188] No vegetation except the wild sage until we came to the river. Mr. Beal,[189] one of our party, walked in advance of the train. We left the road to camp and he was lost. Mr. Otzman[190] and I went in search of him [and were] out until dark and were unsuccessful. Supposed he stopped with some party all night.[191] The wolves howled all night. Gathered a few gooseberries and currants.[192] Travelled about ten miles.

1 July – Left La Prele Creek and passed a small creek of good water and went on to Fouche Boise River. This stream is made by one large spring five miles above.[193] We are now at an altitude of [194] feet. We made only eight miles today. When in camp, a person from another camp came to see if any of us would go to a shooting match for a beef. Some of our party went and won three-quarters of the cow. Mr. Otzman and Mr. Mills went in search of Mr. Beal and had not returned at dark. [There were] a lot of Indians at a short distance from camp. Gooseberries and currants, also mint.[195] A barren country with wild sage and thyme.[196]

[Book Three of *Journal*]

2 July 1850 – Left Fouche Boise River, latitude 42° 51' 5". Four miles brought us to the North Fork of [the] Platte River. Five miles more brought us to Deer Creek, altitude 4,864 feet;[197] swift current with clear water and [an] abundance of fish. A coal mine. Another of our party went out in search of the lost ones and returned with them. We were very glad to see them. They had been treated very hospitably by other companies where they stayed. There are a great many beaver on and about the stream. They are a very curious and industrious community and they build large dams across the streams. They carry the soil upon their tails which are flat, hard and scaly. They also cut down large trees. I have seen a tree over eighteen inches through that they had eaten off and thrown across the stream to commence a dam. I also saw numbers of smaller ones. They display a great deal of ingenuity in the construction of their houses by having different apartments. One of our party shot one last night. They are of a dark brown color with their hind feet webbed like a duck. They have tremendous teeth and very powerful jaws and have a musky smell. When the Indians fell a tree across any stream for the purpose of a bridge they will eat it off in a night. They allow no improvements but their own.[198] There are a great many bears[199] in this part of the country; also elk, deer and other game. Went up the creek to encamp. Made nine miles. Tolerably good grass, beautiful water.

3 July – Last night we had a public meeting of the company to settle some grievances and define what was every man's duty, there having been a great deal of unpleasantness through misunderstandings. We appointed teamsters, loose cattle drivers, hewers of wood, and carriers of water. I am one of the latter. Also cooks. I am also a morning cook and bake every other night. A steward was appointed, the guard reset, and a general overhauling of things. We commenced our new duties this morning. When we were about ready to start some of the cattle were missing. I have been suffering from the toothache ever since I left Fort Laramie so I came across a dentist and he soon eased me by extracting it. He was in a train that had a portable blacksmith shop along and were fixing up a stall to shoe oxen in. The bellows were fixed under the fire or forge which was a kind of table [with a] sheet iron top. A pipe with a curve brought the wind into the fire. A very good arrangement. He had his anvil fixed on a washing tub upside

down and his vise was put in the ground.[200] We this day travelled to Crooked Muddy Creek which deserves its name.[201] On our way there we touched on the bank of the Platte.[202] There were some people crossing on a kind of raft made of three canoes side by side and saplings laid across, but it was so slow and dangerous and likely to create sickness as we should have to be wading in the river at least two days, so we concluded to go to a Mormon ferry about twenty miles further up. We had to go up the Crooked Muddy about five miles before we could find feed for the cattle and it was but poor after all. I should have mentioned before that we frequently go from three to five miles before we can camp, the grass being eaten up by former trains. We made about ten and a half miles[203] and encamped on Crooked Muddy where the people were crossing the river today. There was the body of a man lying on the side of the river. He had on a pair of buckskin pants and jeans[204] coat. He had been dead some time. We had some buffalo steaks for supper and I never ate better meat in my life. In fact it is not the buffalo but the bison. The taking of it was quite an exciting scene. It attacked the horsemen and they had to run for it.[205] They have a thick mat of hair on the front of their heads, also down the back of their legs. It is a chance if a bullet will pass through the hair into their skull. One shot an animal they call the hare.[206]

4 July – Anniversary of American Independence. This day seventy-four years ago this great republic was formed, an asylum for all the poor unfortunates of Europe where they may live and enjoy themselves free from tyranny and oppression. We travelled until three o'clock then laid by on a beautiful clear little creek. We had to let the wagons down with ropes, it was so steep and stony. We intended to keep the Fourth up but it was so late before we got into camp we put it off until tomorrow. At night the young folks got up a dance and singing.

5 July – Remained in camp all day so that the cattle could recruit,[207] there being good grass. We also killed a large calf and had a regular barbecue.[208] The ladies made a lot of pies and we all ate dinner together. The first good meal I have partaken of since I left home but unfortunately just as dinner was over or in fact scarcely, one of those mountain storms came on, which come so unexpectedly in this region of the country, and for a short time spoiled our sport. There was a general rush towards tents and wagons; the wind and hail lasted a short time and afterwards it was a pleasant evening. The band played several

national airs and all were good humored and those little jealousies were at least for a time forgotten. Some shot at a target, others took a game at whist and some sang. I had a tremendous washing and mending which occupied the whole of the morning and part of the afternoon. One party went a hunting on the mountains but were not very successful.[209] They saw mountain sheep,[210] etc., but [found them] hard to get at. They brought back ice and one handkerchief full of snow so we snowballed one another on the fifth of July. We can see some snow on the mountains. There was a hail storm when they were up [in] the mountains and the pieces of ice that fell were incredible. We fixed up tables with tent poles and forked sticks and put some wagons' boards across and under the shade of some box elder[211] and bitter cottonwood trees.[212] The creek ran round our dinner table in the form of a horseshoe. We found a beautiful spring of water.

6 July – Started for the ferry[213] but found a fixing some emigrants had made, so we bought them out and commenced to ferry by taking off the wheels and bodies. We ferried over in two wagon bodies.[214] Had a rope fixed across the stream and worked ourselves over so we got over ten wagons and it then became dark so [we] quit until the morning. Made only about five miles.[215]

7 July – This morning [we] got our breakfast and commenced ferrying again. We came very near losing some of our party as they were taking over a wagon body. The boat filled with water and they were thrown into the river but two held on to the rope that crossed the stream and so worked themselves out.[216] One could not swim and when he came to the bank he was quite exhausted. The other went down the stream with the wagon bodies.[217] He could not swim but held on and the other boat went after it and it was gotten on shore about two miles downstream, all safe. We got all over by ten o'clock.[218] We had to make the cattle swim over. A few wrote home but I had no opportunity, having to repair a wagon that had been broken yesterday morning and then to make a new table frame. One of the ladies, a Mrs. Dexter, the wife of our guide, was very sick so we had to stay in camp.[219] We crossed several teams over the river at $2.50 per team.[220] We are getting a very high wind today. The owners of the ferry above here have sold out and returned to the States having made $50,000 by their speculation.

8 July – Remained in camp on account of Mrs. Dexter's sickness which was brought on by fright. The wagon that went down the river belonged to her husband. Twenty-one of the cattle [were] lost.[221] They

being on an island, we had to swim to them. It is very monotonous staying in camp. We try all kind[s] of things to kill time. Some fish; I have been busy with music. I remembered one of the old tunes the Ellesmere[222] Band used to play so one of our musicians put it on paper and we have been trying to play it. Found eleven head of the lost stock.[223] A tremendous high wind blew down the tents and tore some to pieces. Cool night.

9 July – Went in search of the lost cattle. Great dissatisfaction among the men on account of crossing at this place instead of the regular ferry.[224] Saving a few dollars and losing several cattle beside losing our time which is more valuable than all. Found the cattle; all but two cows.[225] Concluded to start and went up the river a piece to fetch the cattle. Saw a grave where the wolves had scratched up the body and eaten it up. The bones and remains of clothing were scattered about the plain. If his friends could have seen it, what a sight for them. The wolves howl round the camp every night but are afraid of our fires. We returned with the cattle and left a few to bring the horses. They had to be crossed off an island and the stream ran very rapidly. We had just arrived in camp and were beginning to yoke up when one of the party came in on the gallop and said that one of the party was drowned: William Faulkner, a very fine young man.[226] He was a native of Lancashire and his uncle and cousin were of the party.[227] He could not swim and undertook to swim over on one of the horses but he was washed off the horse's back by the current and the horse trampled him under. Several swam in but too late. He raised once and then about a hundred yards below he was seen again and that was the last of the poor fellow. It threw a gloom over the whole camp. His only brother was drowned in England. His sister will be very much affected when she hears, she thought so much of him. So far this month has been full of distress. We made two miles and camped in a grove of cottonwood trees.

10 July – We left the Platte River I hope forever. Our road lay over a barren, sandy waste. The wind being very high, the sand was in perfect clouds and looked like waves the way it rolled along. We passed over Red Buttes, hills of red sandstone.[228] We travelled about fifteen miles and encamped on a mineral spring or creek where we had no fuel but the wild sage and we brought the water with us.[229]

11 July – The mineral spring is not good to use; [it is] considered poisonous. No bad taste unless the cattle trample in it. In that case it

becomes black and is doubtless poisonous. We also passed Rock Avenue,[230] a steep descent. The road here passes between high rocks forming a kind of gateway for a quarter of a mile. Two miles brought us to the alkali swamps and springs. They smell bad and we guarded them while the cattle passed. We next came to Willow Spring; most excellent water.[231] We filled our kegs and went on our way to Prospect Hill.[232] Here we saw the Sweetwater Mountains[233] and miles and miles of barren, sandy plains and hills, no grass, nothing but the wild sage that appeared as if it did not care about growing. During our drive today I counted thirty-five dead oxen,[234] four horses, and one calf lying in the road; some that had drunk at the alkali swamps and others overdriven in attempting to cross this desert quickly to get to grass. We encamped on a small stream running into the Sweetwater. Warm and windy. On watch.[235]

12 July – Slight frost during the night. Made an early start and came to the greasewood (Absynthea)[236] where the sage had attained the height of from four to five feet and the stems were as thick as a man's thigh. Road very sandy and several of the oxen gave out and we left them as prey to the wolves.[237] We arrived at the alkali springs and lakes where persons who are short of lalescitors[238] (carbonate of soda) for raising their bread can supply themselves. It is on the top of the swamps in white cakes. The swamps smell bad; [their] water is poisonous. A few miles further brought us to the Sweetwater River,[239] a very pretty stream of clear water. We ate dinner and then moved six[240] miles out of our way to feed and recruit our cattle. Found tolerably good grass. The wild sage grows here from eight to ten feet high and a foot through.[241] I could not have believed it if I had not seen it. Several elk came close to camp. Travelled about twenty miles.[242] Very tired.

13 July – Remained in camp all day. Some went hunting.[243] We had to repair several of the wheels. They were very loose so we had to fix them as before. An unoccupied spectator who could have beheld our camp today would think it a singular spectacle. The hunters returning with their game; of the women some were washing, some ironing, some baking, others sewing, etc. At one of the tents the fiddle and flute were sending forth their melody among the solitudes of the Sweetwater. At one tent I heard singing. Some reading novels and another her Bible. While all this was going on, that nothing may be wanting to complete the scene, a Negro woman was singing hymns. Others playing cards, so you see we have a miniature world among

ourselves. A mixture of good and evil, which shows that the likeness is a true one. Performed a great tailoring job; the knees of a pair of pants becoming thin, I cut off the legs and turned the back to the front. As we expect cold weather soon I [would] like to be ready for it. Fixed twenty-five wheels. Very warm today.

14 July – Frosty morning and the hot coffee went well. After a drive of five miles we arrived at Independence Rock.[244] This is a pile of gray granite rock alone in an open plain, about one half mile long and one quarter mile wide and about one hundred feet high above the plains.[245] I clambered up to the top on the northeastern side.[246] Portions of it are covered with inscriptions of the names of travelers with the dates of their arrival, some carved, some in black paint, others in red. The Sweetwater River which heads in[to] the Wind River Mountains runs along the southern side and leaves a strip of some twenty or thirty feet of grassy plain between the base of the rock and the river. Five and a half miles brought us to the gap where the road passes through the mountain. About three-quarters [of a] mile to the west there is a chasm or dalles in the mountain called the Devil's Gate.[247] The Sweetwater breaks through this canyon which is from four hundred to five hundred feet high. On the south side the rocks project over the stream but on the north it slopes back a little. The whole mountain is a mass of gray granite rock, destitute of vegetation save an occasional scrubby cedar or pine. From where the river enters to where it emerges from the canyon is about three quarters of a mile. The water rushes through like a torrent, dashing among the huge walls of rock which have fallen from above and they are worn quite smooth. Three of us went through the canyon. By stripping ourselves and crossing the torrent from rock to rock we managed to get through.[248] I never beheld such a grand, terrific, and terrible sight in my life. Although the day was very warm, in the canyon it was quite cold and [it] had a splendid echo. We are now in a valley which reaches to the southwest as far as the eye can reach and the rocks rise on each side of us, mountains high. On Independence Rock where was no vegetation except a few scrubby trees which bore a red currant which tasted like the leaves more than [the] fruit and was very astringent. We found gooseberries in Devil's Gate. We travelled about twelve miles and encamped on the river.[249] Grass tolerable. Just as we were preparing to start, [in] a wagon that was camped close[250] to ours called the Wild Hunters Lone Star which signified they were from Texas, two of the men, one a cripple, had a

quarrel. The sound man was to blame but he drew a pistol and snapped it at the cripple but it happened not to go off. The other whirled round and shot him in the thigh and then rode off and the other after him, but the cripple getting his rifle the other went away, and at Independence Rock he met with a surgeon who extracted the ball.[251] A little snow to be seen on the high rocks. Travelled about sixteen miles.

15 July – This morning ten of the horses were lost;[252] [we] found them, but made a late start. Travelled along the side of the Rattle Snake Mountains;[253] they are composed of granite and trap rock;[254] some peaks high. The river here runs through them but nothing to compare to Devil's Gate; the road here passes through the rock over a barren, sandy country. The wind was dreadfully dusty and blew tremendously and the rain [was] together with the sand. It has been a very unpleasant day; had to ford the river to get fuel. A man shot a hoop snake,[255] the first I ever saw. We encamped on the river close at the foot of [the] Rattle Snake Mountains.[256] Several went up to the top and had difficulty in getting down. [There was] water in holes in the rock at the top of the mountains. Made about twelve miles.[257] One ox gave out.

16 July – Started early and made twelve miles.[258] Saw a lake dried up and encrusted over with a composition of saltpeter and saleratus;[259] upwards of fifty acres. Camped on the river and had to wade for fuel. The roads, where they run close to the rocks, the sand is blown off and it resembles the asphalt or asphaltum they use in Liverpool for roads: hard and black. Barren, sandy roads. There are many bears among the rocks.

17 July – Started very early as we have to make a long drive today on account of water. We passed through the narrows where the river passes through the rocks and the road has to go in the river for some distance. We this day forded the river five times. After we diverged from the narrows we passed over a barren, sandy, rocky road until we came to a rocky pass [where] we beheld the long-looked-for snow-capped[260] mountains.[261] They were the Wind River Mountains.[262] On our right a mass of naked rock, on our left a range of high mountains mostly covered with timber, and in the valley no vegetation. Carried water for fear we could not get through to any. We managed to get through by eleven[263] o'clock, having left the river for 16½ miles, making the day's journey 27½ miles. One ox gave out. It is astonishing how much better the oxen travel in the night. We camped

on the river. The road was strewn with dead cattle and the stench was very disagreeable. There have been more or less dead cattle every day since I took the trouble to count them on the eleventh. We came to Ice Spring, a swampy piece of ground, where by digging down two feet ice can be obtained but [it is] unfit for use, the swamp being alkali.[264] Slight frost. Moonlight. Sandy and dirty.

18 July – Remained in camp until twelve o'clock to recruit the cattle and then went about five miles and found good grass.[265] Crossed the river several times. Cold night.

19 July – Travelled today over rocky hills; some places I thought it would be impossible for wagons to go but we got over with but one accident. One wagon upset containing a mother and child but they escaped unhurt.[266] We passed several creeks, also springs. The snowcapped mountains are on our right and they present a most grand appearance. Huge masses of ice and snow piled up peak upon peak, with large bodies of pine covering parts of the mountains. The country between us and the mountains, several miles, is rolling and apparently barren. The snow is within two miles of us and we have just had some. This has been the warmest day since I left Fort Laramie, but cold tonight.[267] We travelled about seventeen miles and encamped on a fork of the Sweetwater. Our fuel tonight is willows. Rocky hills round the camp.

20 July – As the cattle last night had to be driven some three miles up the creek so as to find feed, the guard had to take a tent with them and watch.[268] The wolves were very noisy round camp all night. As we went up the creek this morning after the cattle, we passed over snow three to six feet deep but frozen hard. Drove the cattle some distance up into the mountain, let them feed and then brought them back to camp and by one o'clock we started. We yoked up some young heifers that had never been worked and it was a difficult job to do. We travelled about seven[269] miles and encamped on the Sweetwater for the last time, the snowy mountains still on our right. Cold evening.

21 July – Last night just at bedtime a severe storm came in so short a time [that] we were quite unprepared for it. The wind blew tremendously and swept tents and wagon covers all before it, and [there was] a fire that was some distance off. The sparks were blown among the wagons and such a scene of consternation; some running one way, some another. The sparks blew into our wagon and we were very much afraid as we had fifty pounds of gunpowder in it. We put out the fire, no damage done, but it blew and hailed all night.[270] Very cold morning

and a dense fog. Looks very like snow. We left the Sweetwater for the last time. We passed the Twin Mounds[271] and ate dinner the other side of them. As we were starting off we were called back. A Mrs. Bolton who had been a little sick for a few days had just died suddenly. She had only a short time before said she felt better. She was aunt to the young man who was drowned. Disease of the lungs, aged forty-seven. We being at such an altitude the air is very much rarefied and we all feel the effects of it more or less. It appears we sooner get out of health than at other times. We this day passed over the dividing ridge which separates the waters flowing into the Atlantic from those which find their way into the Pacific Ocean. We had reached the summit of the Rocky Mountains, altitude 7,489 feet.[272] Six miles brought us to the Pacific Springs,[273] the water of which runs into the Green River or the Great Colorado of the West.[274] We were [now] in Oregon Territory. We encamped and made a coffin out of a wagon bed and I engraved her death, etc., on a board.[275] Made fourteen miles. Very cool day. A misunderstanding having taken place between some of the parties a division took place and our train was broken up.[276] Mr. Wilsey and Dexter, Pullyblank, and Williams[277] left; also Mr. M. O'Neal.[278] We were then only seven wagons, Mr. Bolton having a wagon but no team. Messrs. Beal and Otzman bought three yoke of steers and took their share of provisions and joined the other party; what we all would do if we could.[279] We are now reduced to six wagons and I hated parting with some of the parties, but we shall all be glad to leave our leader Mr. Sweasey.[280]

22 July – The funeral took place this morning. The parties from the wagons who had left also were there,[281] as were many strangers. Some people from Birmingham, England, watchmakers, also were there. We came to a dry creek where we could get water by digging a foot but it was not good. A white sediment, the same as I have noticed elsewhere, covered the surface of the ground. Ten miles brought us to Little Sandy[282] which we reached at ten o'clock in the night having travelled twenty miles.[283] The cattle had nothing to eat since morning and there is no feed here. I was on guard. Cold night; rained more or less all day. Even the sage here has a withered appearance. We are now out of the range of the buffalo which, although not often mentioned, there have been many seen. Little Sandy takes its rise in the Wind River Mountains, still in sight.

23 July – We took breakfast at four o'clock this morning[284] and forded Little Sandy. Went to Big Sandy.[285] They are forks of [the] Green River. Six miles, then we went about six[286] more down the stream for feed. Found good grass. Saw antelope. Frémont's[287] Peak, the highest peak

[Book Four of *Journal*]

of the Wind River Mountains, is about 13,570 feet high,[288] is covered with snow, and has a singular appearance.[289]

24 July – Remained in camp until three o'clock.[290] Cut grass and carried it in sacks to the teams. About two miles out hunters shot some game; sage fowl and hares.[291] As we were cutting the grass three antelope ran and stopped close by but we had no guns and the hunters were away. They are a beautiful animal. Part of us went down last night with a tent and stayed all night to watch the cattle.[292] We have to cross a corner of the desert and we shall travel all night. Had gooseberry pies. We started at three o'clock in the evening, took supper at sundown and travelled until two o'clock when we changed drivers.

25 July – Went on until four o'clock when we took breakfast and fed the grass to the cattle, then rolled on to [the] Green River where we arrived at three o'clock. The first part of the road was good but the last was very bad. We had to lock both wheels and take all the oxen off but one yoke and then slide down hills. Hundreds of dead cattle all along the road; we left one. We swam our cattle over the river and four of us took them four miles up the river and watched them all night. There is a ferry over [the] Green River; five dollars per wagon.[293] This savors of California prices. The distance we have travelled, said to be forty miles, proved to be fifty-three.[294] The wagons cannot cross tonight as the rope of the ferry is broken and they are repairing it. This stream runs at a fearful rate and it washed some of the cattle down the stream more than a mile.

26 July – The watch being relieved[295] we returned to the ferry where they were crossing over the last wagon. Got all over safe. The part of the company who left us are here. This river runs through a wild, rocky country; no wood or wild sage within two miles. We remained here the rest of the day. Dead cattle round the camp – sweet very.

27 July – We greased the wagon wheels and got all ready prepara-
tory to starting but we lost three steers. Started at eleven o'clock.[296] As
we were eating diner there came a lot of Indians of the Shoshones or
Snakes.[297] This is their country. Some of them were dressed in a fine
showy way with brass and tin ornaments and their horses were
trimmed over with trappings of buckskin and their ornaments jingled,
in particular the squaw, but some of the men had on some of the
white men's clothes and those put on so curiously. A vest buttoned up
behind; one had a Spanish saddle turned wrong end foremost, and one
had an old umbrella point with his hair sticking through it. Some had
some old hats on they had picked up; one had a pair of spectacles tied
round his hat. Another had a half-military blue coat and naked legs
with an old sword. Altogether they were a rum[298] lot but they are
excellent horsemen. They are called the Arabs of the Rocky
Mountains. Children very small can ride well. They saluted us with a
smile and held out their hands and said "how du du, how du du," then
laughed. We examined the make of their saddles, their bows and arrows, etc. They are very fond of bead and articles they get but little of.
One wanted powder, as some had rifles. He held up his powder horn
and said "me white mans brudder," then made signs on the ground to
his men. He wanted it to shoot game and not the white men. Some
traded with them for some small articles and they went their way, I
dare say making as curious remarks on us as we did of them.[299] We
made about twelve[300] miles and encamped on a small fork of [the]
Green River abounding in trout, but black spots instead of red.[301]
Warm day; very cold evening with a thunder shower.

28 July – Frosty morning with ice. Started early and travelled over
a very hilly country and saw lots of Indians. Came to one of their villages where there were several hundred of them[302] and we were com-
pletely surrounded by them. We traded for moccasins. They were very
civil and they would hold out their hands to us [at] fifty yards' distance.
They all have the "how du du." Met one who carried an umbrella over
his head although it was dry and he could not spoil his complexion.
One, a chief, very gaily dressed and a fine looking fellow made signs on
the ground that this was their country. They wanted tribute for us
passing through. The squaws are very fond of blue and red and they
make their legging of blue with red tucks. Petticoats are out of
fashion. Their dogs are half-bred wolves and they have lots of wolves
tamed. I saw some of the young ones shoot with their bows and arrows.

33

I tried to trade for a pony but did not succeed. I am regularly tired of walking. We travelled over twenty miles and encamped at the foot of a hill which at some remote period had been an oyster[303] bank. We found shells more than a foot in length and two inches thick.[304] One of these monsters must have been a mouthful. The hill is about seven thousand feet above the sea. There is a great scope for the geologist in these mountains. Kept a strict watch.

29 July – Cattle all safe [and we] started on over ascent of the dividing ridge which separates the waters of the Pacific from those of the Great Basin or desert. We have the worst roads today we ever had; up hills of the longest, steepest,[305] and rocky [kind]. This distance from [the] Green to [the] Bear River is only ten[306] miles in a direct line yet we have to go sixty-four to get there. Made Ham's Fork[307] of [the] Green River by twelve o'clock. As it is eighteen miles to grass we concluded to stay until morning for the cattle to recruit. We were surrounded by Indians. They are great beggars. Had to watch them; they will steal. I traded some gunpowder for a pair of moccasins; as my feet are very sore they will relieve me. Saw an Indian conveyance and a lame half breed, half French,[308] [who] could speak English well, was riding in it. It was just two poles about sixteen feet long brought up each side of the horse like shafts, the other end trailing on the ground; then two pieces of wood tied across and a piece of rawhide put across the pony's back answered for a backhand.[309] Rawhide [strips] were then worked across from pole to pole and a buffalo robe laid on them. The man sat inside with a little papoose child and the squaw sat astride the horse with her feet on the shafts and drove her lord about. The squaws do all [the] drudgery.[310] Travelled twelve miles and encamped on Ham's Fork.

30 July – I had the guard from twelve until four. Frosty ice one quarter inch thick. Passed over the dividing ridge.[311] Saw a spring at the top of the mountain. Camped on a small stream, a tributary of [the] Bear River, in a glen. Made sixteen miles.[312]

31 July – Froze the milk. Climbed a very high, steep hill and descended into the Bear River valley. The bottom is wide and covered with good grass. Crossed Smith's Fork[313] and encamped on [the] Bear River.[314] Made fifteen miles. Very cold night frost.[315]

Idaho

Snake River

● Fort Hall

4 August
Soda Springs

12 August

Raft River

Bear River

Wyoming

City of Rocks

Nevada   16 August

31 July

Utah

August – Warmer morning than usual and lots of mosquitos. A great quantity of yellow and red currants but very sour. Bottom very wide and covered with grass. Camped on Thomas's Fork.[316] Made fifteen miles. Went a[317] fishing.

2 August – Had a delicious mess of fish for breakfast; suckers,[318] trout and rock bass.[319] We passed over a mountain called the Big Hill.[320] We got over safe. The ascent was three-quarters of a mile to [the] top and very steep. The road then runs through a ravine where there is just room for the wagons to pass between the rocks. Then to descend it took a long time, it being one of the worst places on the whole route[321] and glad we were when safe at bottom; we descended into the Bear River valley close to Bear Lake; a large lake covered with wild fowl. A high range of snow covered mountains [is] to be seen to the northwest. Swam across the river to fetch the cattle. A cold frosty morning. There is a large quantity of large crickets, some three inches

long.[322] When killed, the others will eat them directly. We encamped on Spring Branch[323] coming in from the mountains. Lots of mosquitos. Green scrubby willows for fuel. Good grass. After supper we found a dead ox in the creek above our camp. Delightful very. Made twenty-two miles.[324]

3 August – Made twenty-two miles.[325] Good road. Had some Indians shooting at biscuits on the top of a stick. When they shoot one they eat it. Some good shooting with bows and arrows.[326] Encamped on [the] Bear River. A great many of the rocks about here are of volcanic origin; some a kind of basalt. Caught fish.[327]

4 August – Two of us started off to reach the soda springs[328] about five miles [distant] but found them about ten miles distant. The first view we had was two white mounds or hills on the right of the road. One is about ten rods long and four in width[329] and about thirty feet high. The size of these mounds continually increases as the water oozes out at different places, which becomes a hard crust. The rocks for miles around are of the soda formation. Upon these mounds the water is warm. About two hundred yards from these mounds is a hole in the ground in the solid rock about two feet across the top and wider at the bottom like the inside of a jug. The water is bubbling and sparkling and strongly impregnated with soda. It would sometimes swell four inches above the surface. This and several others near afforded good drinking water. It was cool and when sweetened would compare favorably with any soda water I ever drank. Just below the mounds and by the side of a grove of cypress[330] and pine is a rapid stream formed by the soda springs. Where the road crosses the creek there is in the bank side a hole in the rock where it spouts through in a stream. That was the best to drink. Very odd. After crossing the creek, in about three-quarters of a mile the springs boil up in every direction. Several mounds have been formed ten feet high. The water has found some other passage and left them to molder away. The center of these are concave. The bank of the river is rock of the soda formation. The space between the river and the mountains is barren. The soda has left a sediment, crinkled and loose, and some few mains, but no water runs from them now. The river here is about 150 yards in width and about two feet deep, running rapidly over a rocky bottom. The soda water bubbles up in every direction and sometimes rises above the surface of the river six or eight inches. This bubbling extends one-half of a mile. Here there is a small stream coming in from the north which tumbles

over rocks and falls into the river. Where the road crosses there is a circular basin in the rock about two feet across but larger below. It was covered with grass; the water every few seconds boiled over the top and we happened to see it. On that account very cool and remarkably clear. In some of the small holes the water is red and rusty, being strongly impregnated with iron and having a very sulfur taste. About three hundred yards further down the river is what is called the Steamboat Springs.[331] The water has formed a small cone of about two and one-half feet in height and three feet in diameter at the base. A hole of six inches diameter in the top allows the water to discharge itself. It swells out in disjointed fragments every eight or ten seconds to the height of five feet and it is warm and has a milky appearance but when taken in a glass is as clear as crystal. It provides a sound similar to a steamboat but not so loud. About six feet from this is a small fissure in the rock which is called the escape or gas pipe. It makes a hissing noise. The gas from this would suffocate a person if he held his head to the ground for a short time. To the right of this there are several mounds sixty or seventy feet high but nearly dry and falling to decay, and the rock is in powder of various colors; white, red, yellow, orange, pink, green, and all the intermediate shades, and [I] expect [it] would make good colors for painting purposes. The water in many of the springs is strong enough to raise bread like yeast. Providing these springs were near any settled country, they would be much resorted to in the summer months as this part of the country appears to be perfectly healthy. I should like no better fortune than to possess one of them in New Orleans.[332] Five miles brought us to where the road leaves the river and bears northward through a valley.[333] The river bears southward and empties into the Great Salt Lake. As the river turns, the point of the mountain is composed of huge masses of basaltic rock several hundred feet high.[334] The river runs through the rocks, dashing and foaming at a fearful rate. From where the road leaves the river it is very rough, being over masses of lava and basaltic rock. Craters are yet standing in the plain of volcanic origin; one a large hill with the mouth of the crater at [the] top and hollow with blackened walls about three hundred feet across.[335] Around the volcano there are pieces of pumice stone, cinders, lava, and pieces which are melted together like black glass bottles broken. They lay scattered in every direction and [there are] tremendous fissures in the rocks; some are fifty feet deep. Others have been raised out of the ground and left in that manner so that they present a regular breastwork, as if to fortify an army. We went on across the valley and got benighted without any water and went to bed supperless.[336] After running about all day as I had done to see all I could, I was very much tired and soon fell asleep. We made at least twenty-five miles.[337]

5 August – I was awakened by the ringing of the bell. A very cold morning. Yoked up and off we started by five o'clock to try for water. Found a paper by the road side saying fourteen miles to water. That was rather staggering but we could do nothing but go on so in about eight miles we came to a beautiful stream.[338] Got breakfast and a good wash and felt quite revived. Rested awhile. Sun very hot in the middle of the day. Feet sore. Travelled about fourteen miles.[339] Had to climb over a tremendous hill and had to double teams. A chain of one wagon breaking, the wagon went to the bottom; no harm done. Had to fasten ropes to the sides of the wagons to keep them from turning over. Encamped on a mountain stream of cold water. Saw elk.[340]

6 August – We this day made twelve miles and took dinner by the side of a swampy marsh where we shot some ducks and geese.[341] Went a little further and encamped in a beautiful valley covered with good grass. Very hot. Passed a cascade on a tributary of the Columbia.[342] A beautiful stream.[343]

7 August[344] – Ducks for breakfast, then went shooting in the swamps and the wagons went on. Found the skeleton of a human being in the swamp.[345] Had been there a long time. Overtook them [the wagons] by dinner time. Very hot day. Passed over a small mountain with a spring of water nearly at the top where someone had been amusing themselves by erecting a miniature mill wheel. We left it grinding away. We came to [the] bottom of the hill and one of our best steers fell down dead in less than five minutes. Shortened our journey by twelve miles. An Englishman from Liverpool came by us last night. Morrison by name. His father was formerly a builder there in Toxteth Park.[346] He was alone with his family so we told him to join us which he did.[347]

8[348] August – Met several Indians and they became very trouble-some. Made about sixteen miles. When we were encamped several Indians came. They are great beggars. We gave them a large iron pot to eat out of. There had been porridge made in it. They also made signs for some fat. We gave them some bacon fat in it. They laughed and chattered and the way they cleaned the pot, no dog could have made a cleaner job of it. They were Shoshones.[349] We are nearly out of their country. No fire wood. We found some wagon wheels, so cooked supper with them. I wish I could have taken a sketch of the Indians as they were sitting round the pot.[350]

9[351] August – Made a late start on account of losing a steer. Found it, then travelled twenty[352] miles without water. Went up a ravine and hill six miles in length and down a regular precipice of a hill where we locked the four wheels at once. About a mile down. Very hot. Encamped on a small creek.

9[353] August – Travelled about eighteen miles and found no water. Encamped and then found a small spring about two and one-half miles distant but no water for the stock. Tolerable grass and very warm day.

10 August – Very warm night last night. Travelled over a hilly country and entered a ravine, how far through we did not know,[354] so we travelled a long way and still the rocks were on each side of us and just room for a wagon to pass. Could see nothing but the blue sky overhead. The hills and rocks on each side were several hundred feet high. Came to a hole where someone had dug; a little water in it, but it was not good. Gave a little to the cattle but they did not seem to care about it. We then went on, thinking to make the end of the ravine in an hour or so, but we travelled on until dusk, when Messrs. Sweasey, Bennett, and myself went on ahead to see how far it was and came back to encourage the others. We went about six miles and the end appeared as far off as ever for we could see nothing but rocks and a little sky. We then sat down and rested and waited for the teams, but they did not make their appearance. We began to get alarmed that something had befallen them so we turned back and in about three[355] miles we came up to them. They were unable to proceed further on account of the darkness and dust. The cattle were tied to the wagon wheels and the men in bed, having made nineteen miles. Very warm. A singular circumstance occurred; two currents of wind at the same time from different directions; one was cold very and the other was warm to suffocation. Very close night. Nineteen miles.

11 August – Started off early without breakfast. After travelling about six[356] miles we made a small stream running down one side of the ravine so we got breakfast and the cattle some water. We then went on until we came to the end of the ravine. It proved to be at least fifteen miles through. We emerged on to a flat, open country with

mountains in the distance. We thought we saw a creek a few miles distant but when we arrived at the place [we found we were] mistaken so we went on and the cattle were very much tired. Some of them laid down. About ten o'clock at night we made a creek and took dinner and supper together having made twenty-four miles. I was very tired and footsore.[357]

12 August – By the daylight this morning we were enabled to see what kind of a place we had got into. Tolerably good grass, but the creek [was] full of dead animals. Although not often spoken of, the dead animals will average over thirty per day, one of our party having kept an account, but we see very few graves. We are now on the headwaters of [the] Raft River,[358] having crossed over the dividing range of mountains between the Columbia River and the Great Basin. During breakfast this morning a small bird of a brownish color with its feathers fringed with drab[359] came into our camp. It was so tame it jumped on our shoulders, went into the tents and made itself quite at home. How it came to be so tame I cannot account for without some emigrants having tamed it. At any rate it was silly to repose so much confidence in man. It escaped unhurt but the next time it ventures may be its last. We fed it and when we were leaving it jumped on the tops of the wagons. Some wanted to take it along but it would have been very cruel to deprive it of its liberty, a thing dear to us all, when it put so much confidence in us. Travelled seven miles and encamped on the headwaters of [the] Raft River at the junction of the Fort Hall road. We saved about thirty miles by coming this cutoff.[360] We saw some Germans leaving their wagon on account of losing one of their mules. They started to pack as a many have done before but they were mean enough to set fire to their wagon before they left it. That was too much like the dog in the manger, because someone might have needed one. If they needed it to cook with it, [it] would have been different.

13 August – Crossed over the creek and started on the regular road. We have now again lots of company. Met two men; packers with two horses. They had been round to the Salt Lake City of the Mormons. They were for Oregon. They gave the Salt Lake valley and city a great name for its fertility and pleasant location. It is quite a large city already. It is situated five miles back from the lake. They appear to be very busy carrying on a great trade with the emigrants in buying their broken-down cattle and trading them theirs. They then

40

turn them out on the grass which grows in great abundance and they are ready for sale in a few weeks. They also carry on a considerable trade with the Indians. After dinner we left the creek, filled our water casks, passed a barren mountain,[361] and encamped on a beautiful piece of grass but no water. Made twelve miles.[362] The wild asp[363] from six to eight feet high, silver oats,[364] and flax[365] flourish here in great abundance, also a kind of grass eight and ten feet high.[366] The cattle like all this.

[Book Five of *Journal*]

14 August – Showers and rain. Started early and made two springs of good water and crossed several spring branches. I was very unwell and consequently had to ride in the wagons all day. At dinner time we were in a large and remarkable glen place which I named Pyramid;[367] about two or three miles long and from three-quarters to one mile in width. [The] formation [consisted of] most singularly shaped rocks composed of granite and lava. They stood apart from one another and at a distance had the appearance of a town with castles, towers and chimney stacks, and one had the appearance of a splendid Gothic architecture. They were nothing more than a mass of cumbersome rocks of an unseemly appearance. The glen was surrounded by rocks of the same description and very high. One in particular [was] split at the top and after we had left the glen [it] could be seen at a great distance. We have met with some people who were hunting lost cattle. Their cattle had stampeded and run away several days before. We found three this morning and gave them to them and gave them information of the rest which we had seen before. They were very glad. We also picked up a horse. It was owned[368] in the evening. We are still in the mountains and have very rocky roads. Rain all afternoon. Made twenty-two miles and encamped on a spring. Tolerable grass and they drove the cattle up the side of the mountains. Very little game in this part of the country. Crows[369] and hawks[370] appear to be the only inhabitants of this desolate region.

15 August[371] – Got an early start but a very rough road. Cold morning and very hot middle of [the] day. Took dinner on a creek; the waters of [the] Rattle Snake River.[372] The banks are covered with willows and birch.[373] Our fuel lately has been greasewood; it is warmer than sage and that is needless. Followed along the creek and encamped on [the] Rattle Snake River. Made twelve miles. Shortly after we

arrived in camp we were visited by a long-threatened thunder shower which delayed our supper. Our boss[374] got up wrong end foremost and was in a bad humor all day.

Idaho

16 August

Mary's River

Humboldt River

21 August

Utah

17 September
begin packing

11 September
Carson Sink

Lake
Tahoe

Carson River

19 September

California

Nevada

August – Made a late start but travelled about eighteen miles. Passed through a small canyon one-half mile long. We then went on along the R[attle]. S[nake]. River, or Goose Creek as some call it, and then entered a canyon four miles long. The entrance was bold and beautiful, the rocks overhanging our heads hundreds of feet[375] above us and the creek running along the bottom with room enough to let the wagons pass. We had to cross the stream twice in the canyon. The rocks were very picturesque and a great variety of colors. If an artist had colored them the highest colors they really are he would be accused of unnatural tricks. The rocks were bright as gold, red, green of every shade, yellow, orange, and all intermediate shades. We also passed a hot spring and we encamped on Goose Creek. Had fresh water mussels for supper.[376]

17 August – Cool morning. Took a copy of the Mormon guide from some people who camped near us. They had been to Salt Lake City. By dinner time we arrived at the entrance to the Valley of a Thousand Springs.[377] There were numerous springs which came from under a pile of rocks. Water not very cold and formed a creek which ran down the valley. The stream, after running a few miles, loses itself in the ground. We went on until night and arrived in a large part of the valley where there had been good feed. Wild rye,[378] but it was eaten off so close that it resembled a stubble; but the cattle did tolerably well. We could find no water but dug holes in a dry creek three or four feet deep and obtained water that was better than none.[379] Shortly before our arrival into the valley a little boy (Mr. Sweasey's), six or seven years old,[380] was getting into a wagon and he fell off the tongue and both wheels passed over him, one over his body and the other over his head.[381] It nearly cut his ear off and tore his cheek dreadfully. We made about twenty-four miles and kept a strict guard all night on account of the Indians who we heard had been troublesome here.[382] Cold night.

18 August – The little boy [is] much better than expectation. Stopped at twelve o'clock to bake bread[383] as we had no opportunity last night. There were several trains here at a fine, large spring. They told us that yesterday six Indians came suddenly from the neighboring hills and attacked their cattle with their bows and arrows. They shot three arrows into one ox but did not kill it.[384] The Indians made their escape into the hills. A good, level road all day; quite a treat. Cool wind all day. Our fuel today was wormwood. We have used it several times lately, but it is a very poor makeshift. We heard from a party that came along that the other part of our train was thirty miles behind. They also informed us that they found a body in the Platte which no doubt was poor William Faulkner [who] we lost there. They buried him. Travelled about sixteen miles. As we are out of bacon, indeed meat of any kind, we had to kill one of our cattle.[385] Our flour is barely enough to carry us through. Some are in a dreadful state for the want of provisions. Some have been so hard put that they have cut pieces of the dead cattle along the road and jerked it then took it along. Some have offered any price to us for flour, but we have none to spare. It is

dreadful to have to refuse a person four or five pounds of flour, but it cannot be avoided. One poor Dutchman had his wallet taken from him by the Indians. He was alone, trying to get through with his bit of grub on his back. We keep a very strict watch.[386] One time a poor German woman came crying to our camp. They had no provisions, having lost them by the wagon sinking in a river. We gave her a little cornmeal and a plate of mush which she was truly thankful for. Indeed every day now we see dreadful cases of want and the poor creatures have five hundred miles to go yet. The time appears to pass very slowly now. I suppose it is on account of getting near our journey's end. We passed through a canyon five miles long; a creek ran through it and we crossed and recrossed it nine different times. It was the worst piece of road we have had yet. The wagon wheels at times had to pass over rocks three or four feet high and the wheels on the other side of the wagon were down as low. We were a long time in getting through. There were also hot springs in the canyon but not boiling. They made the creek too warm to drink. We made about twelve miles and encamped in a very large valley where grass of all kinds grow bountifully. The best grass since we left home. I called it the oxen's paradise. A creek runs through this valley, a tributary of the St. Mary's River, or Ogden's as it has been called, but it is now named after that Nestor of scientific travelers, Humboldt River.[387] A snowy mountain to the south. Very hot day and dusty roads.[388] Very cold evening.

19 August – I was on guard last night and it was very cold.[389] After travelling about eight miles we came to the Boiling Springs.[390] They covered about two or three acres and the boiling[391] water from them formed quite a large creek. The air all round them was quite warm and such a steam arose from them it was like being in a large washhouse. Here was a fine chance for washing and cooking. The water was beautifully clear but slightly impregnated with iron. We this day passed the dividing range between Oregon and the Basin.[392] After travelling about twenty-six miles we encamped at a large spring of good water. We passed a large tract of wild rye containing not less than fifteen thousand acres. Good grass but cold night. It was ten o'clock when we got to camp.[393]

20 August – Very cold night and froze ice about one-half inch thick in the buckets. At breakfast we also, by way of diversion, had a small fight today between a bullheaded Englishman, native of Preston,

Lancashire, and a Canadian about driving team, and am glad to say that the colonist whipped John Bull.[394]

21 August – Made a good start. A lot[395] of horses belonging to some packers came by us in a complete stampede; some had lost their packs, others with their packs turned under their belly. We succeeded in staying their wild career[396] and in a short time their owners came along and took them off. Flour has been sold as high as fifty to one hundred dollars per hundredweight. Made twenty miles and en-camped on the Humboldt River.[397] Good grass; lots of fish in this stream. I had the first watch tonight. Cold night.

22 August – Frosty morning. Some of the watch heard Indians about in the night. Saw a party of Indians. They were a dirty, miserable set and had a lot of horses, no doubt stolen, as they were shod. This tribe is named the Diggers on account of their digging after roots for food as they are like other Indians about work and are too lazy to hunt, but are great thieves and dreadful cowards.[398] One had a beautiful mantle of mountain fox skins.[399] At dinner we were visited by three Indians. One wanted a gun mended and we traded some fish hooks for some fish.[400] Most of these wretches are naked, or nearly so. One had a coat and no pants, another a shirt, and so on. I have frequently heard that the North American Indians have no beards. It is not so. One old man had a gray beard, but very thin. The younger ones pluck it out. Saw a party who had three of their horses stolen by the Indians and had only one horse and ox left. They had a pack on the back of the ox. As soon as we arrived in camp a man called on us inquiring after a young man who had stolen two of his horses, but we had not seen him. Made sixteen miles[401] and encamped on the river and then took a swim in it. Very dusty roads and scorching hot in the day.

23 August – Started early. Two of the party who had separated from us, by walking ahead of their train came up with us. They took the Fort Hall road and consequently were longer. They had lost several oxen. Their train did not overtake us.[402] The roads are dreadfully dusty. We look like a lot of millers, only darker at a sight. We crossed the North Fork of the Humboldt River and after travelling about twenty miles we encamped on the river. One of our party bathing saw some hot springs on the other side of the river. Several of us crossed over and all the other boiling springs we have seen before are not to be compared to these. One of them is circular, ten feet in diameter and

five feet deep, boiling up as clear as possible. We caught a snake and threw it in; also a frog.[403] They gave one kick and all was over and as soon as we could get them out they were cooked for the flesh all came off the bone. There are many of the springs and the water forms a stream and enters the river and makes it quite warm. Saw several Indians and one asked very plainly for tobacco.[404] Asked him how far to squaw, wigwam, papoose by signs. He laid his head on his hand as sign of laying down to sleep and then pointed to the sun and drew his hand gently down to the horizon then lifted up both hands with all his fingers strait up which made 10 suns. They were ten days travel off.[405]

24 August – Shortly after starting we were overtaken by Mr. Dexter's party, they having split from the others, they having had wars and rumors of wars in their camp.[406] We passed them again in the evening. Made twenty-two miles. Passed through a rocky canyon five miles long and crossed the river in it four times. One party got all their provisions wet. Encamped on the river. Very dusty roads and hot weather.

25 August – Made an early start. Met with several people who had been the Salt Lake route and in endeavoring to find a nearer route had got into difficulties. They had to cross a ninety mile desert and were attacked by the Indians. They saw the graves of others who had been killed by them. One boy, they scalped him and cut all the flesh off his bones. One party shot several Indians and many such accounts we hear daily. Several of those who crossed the desert gave out and could not get any farther. Others offered five dollars for a drink of water.[407] Some took water hearing of the distress and sold it at one dollar per quart. I expect a few only will try that route again. We travelled twenty-five miles and encamped on the river having been from it for more than seventeen miles. It was dark when we got there and there was no grass. Dreadfully dusty we are, and hoarse. I picked up a human skull as we walked along today. It had evidently been alive this season as the muscles and part of the skin were on the head. Not being a anatomist I could not tell if it belonged to an Indian or white man. At any rate the wolves had picked it tolerably clean.[408] Cloudy and a shower.

26 August – Frost. The cattle getting no feed last night, we had to remain in camp this morning. At twelve o'clock we started and travelled about eight miles.[409] Good road and found good grass. Caught fish. Thieves or Indians were prowling about our camp last night but the dogs ran them off before anyone could get after them. I

expect it was a party of emigrants short of provisions. Still dreadful accounts of the Indians on the Salt Lake route.[410]

27 August – A party of Cherokees from Lewissee[411] passed us today on their way to the "diggings."[412] They are pretty well civilized and dress and act like white people. Had mules. Saw a kind of small cabbage or kale;[413] also tongue grass or cress[414] [and] babsen mint.[415] The sunflower[416] grows all over the bottom. Saw a grave with a board saying that the interred was killed by the Indians on the tenth day of August, age twenty-two from Iowa. There being a fork in the road today and I being a short distance ahead I took the wrong track and the wagons crossed the river. I kept walking on and seeing they did not come I got on a small hill and there they were three miles at least from me and the river between us. I found a suitable place and waded across. I was afraid the Indians might see me and take a shot at me as they always attack the unarmed and few. Travelled twenty miles. Lots of good grass.[417]

28 August – Severe frost. Shortly after starting we overtook Mr. Dexter's party,[418] but at dinner time they got the start of us and hoisted a flag. We fired off a pistol. We are having quite a race. Made eighteen miles and passed over a stretch of desert seventeen miles [wide].[419] Very dusty and hot, not a breath of air.

29 August – Had breakfast by moonlight and started off early. Passed[420] Dexter's [party] and hoisted a flag. One of our musicians played the air "Good-bye." Last Monday[421] morning Messrs. Jackson[422] and Moore[423] stayed behind to get Mr. Moore's pony which he had lent to Mr. Bolton to see after a lot of geological specimens he had left in another train for them to haul, and they have not returned yet. We are afraid some ill has happened to them. Shortly after we had encamped tonight the lost ones returned. They had got on the other side of the river and had been a long way down. They then crossed over and found us. They had been nearly forty[424] hours without food and were nearly starved. One could scarcely walk and they had to hide among the long grass and willows to hide from the Indians. They caught five minnows[425] and ate them. They slept without coats in the grass. We were very glad to see them return. One of them had a small heifer in the drove and I sold it to some packers who were out of provisions for five dollars and he was pleased I had done so.[426] Made twenty miles.

30 August – Frosty morning. Started out and soon overtook Dexter's party. They had lost a horse and a party went in search but were unsuccessful. No doubt stolen. Saw a curious kind of plant, the resin or turpentine plant.[427] The leaves and the ground beneath were covered over with a gluey substance which looked like dew. By putting it in the fire it will make a great blaze. We were crossing over from the road to the river and had to cross a slough[428] and the wagons, all but one, got mired down. We had to double teams to get out. Crossed over some very bad, rough, dusty hills. The Humboldt River has now become a very dirty, slimy, sluggish stream and much deeper. Made twenty-two miles.[429] Several of the cattle got mired in the river in attempting to drink and we had to go in and get them out with ropes. Then all the cattle crossed over at another place so we let them stay until morning. I was on guard. Cold night.

31 August – Slight frost. After breakfast we had to get back the cattle. Two beside myself volunteered to swim across the river and fetch them; so putting on an old pair of pants, in we went, but wasn't it cold. Got them over by digging down the side of the bank. We then wended our way among the hills, the bottom being miry, but so dusty. The ground on the hills is like a brown lime and hurts the eyes very much and when mixed with water it swells like a bed of mortar. It would seem by the appearance of the soil that it would be a matter of impossibility for anything to grow yet there are several curious plants that are very full of water which is a mystery to me where it comes from.[430] Very pleasant day, cool wind, but I had the rheumatism bad. Made eighteen miles, but lost several, having to go a long way round to head a slough. Saw an Indian grave which had been made a few hours before we came along. He was lain on the top of the ground and a few spadefuls of soil thrown on him. He had been shot and the flies were thick about. He had been dead some days.[431] We heard of an Englishman behind who had all his cattle stolen by the Indians but one, and from the description I am afraid it is Mr. Morrison who I mentioned before, he having stayed behind us on account of his wife's sickness. A German waded through the river last evening and begged a bed and his breakfast from us. He was on foot alone and no food. It is truly heartrending; the cry there is for food. Among the curious things we saw today were a numerous tribe of lizards among which was the chameleon.[432] The ground along the edge of the hill on the bottom is encrusted over with a kind of saltpeter, and that bitter.[433]

Encamped on the river. Tolerable grass; cool and looked like rain. The air is filled with a kind of mist like Indian summer. We cannot see far.

1 September[434] – Started early and had a dreadfully heavy road of sand. Some of the oxen gave out and we had to put others in their places. One of our party, Mr. Jackson, started to pack through the rest of the road. He crossed through the river to meet another party that is going with him.[435] He had Mr. Moore's[436] pony and my gun. He took sixteen pounds flour and a little tea and a small bacon which I gave him, having a few pounds of my own. We made fourteen miles and found middling grass.[437]

2 September – This morning we concluded to leave one of our wagons as our team is getting weaker every day, so we distributed the load among the others and left it – the one we traded for at Fort Kearny, No. 2. We have now five [wagons] in the company.[438] We had bad roads, part of it sandy and heavy hauling, part ashes so light our feet and the wagon wheels sank so deep it was very bad to walk and so light the breeze blew it about until we could not see one another. Travelled twelve miles and found good buffalo grass.[439] Killed some snakes, also a beef, having had no meat for several days. It was a heifer we picked up in the Kickapoo nation and so wild that we have always called her the Kickapoo. Yesterday a man came to sell us some beef but we looked at it and saw it was very poor and turned out to be a lame steer we left the same morning. They had killed it.[440]

3 September – Made a late start on account of some of the cattle being lost. As we were starting one of our company saw a stray horse come down a hill. He asked me to go shares. [We] agreed and went to get it when we found two and after taking them twelve miles we found the owner. Poor fellow, he had been out all day in search of them and was very glad to get them. There is a man near us who has packed on the back of a bull all the way from the South Pass and he has had scarce anything but what fish he could catch since he has been on the Humboldt River.[441] We have been offered a dollar a pound for cornmeal but have none to sell. The river now presents quite a different aspect. It is more sluggish and dirty and but poor grass. Saw two graves of men who had been found drowned in the river. Made sixteen miles through dreadfully bad roads, the dust worse and worse.[442] It was two inches thick on our bed clothes in the wagons. Very hot.

4 September – Remained in camp all day to cut grass to carry for feed across the desert. Had to cross the river. Saw an old wagon bed

yesterday about three miles up the river.[443] Went to fetch and bring it down the stream but the stream is so crooked that it took one-half day to fetch it, being fifteen miles by way of the river.[444] Washing and I nearly said ironing but we mangle them instead, but this stocking mending beats all; it is decidedly the worst job. Cooked a lot of bread for the desert. If we have had one person asking to buy food at any price today there have been twenty.[445] Got over the grass all safe. Killed ducks.[446]

5 September – We sold some packers a quarter of beef at one-half cent per pound and they were glad to get it at that. Travelled on until dinner time where we were overtaken by some packers who had two of their horses stolen last night. They wanted us to carry their clothes and offered a horse for payment but we are too heavy laden as it is. Three poor fellows came up and had not tasted food for two days and were getting quite unable to walk. One wanted us to take him in but we were compelled to refuse as we are short of provisions.[447] We gave them an old ox and some salt, also a few beans and very much pleased they were. They killed it and commenced to cut off the meat and jerk it. They have one horse so they will be able to carry enough to do them through. Travelled on over a barren waste up to the ankles in dust and ashes; the whirlwinds blew it about so it was very annoying to our eyes. Although not mentioned before, the whirlwinds happen here every five or ten minutes, blowing dust and sticks and rubbish over us. We are the dirtiest devils every night that it is possible to conceive. Encamped on the river. It is now a very nasty stream and I have not seen a spring of good water for more than two hundred miles. Made eighteen miles.

6 September – Travelled sixteen miles,[448] our printed guides being so very indefinite that we can't tell where we are, but meeting with two Mormons on their return to Salt Lake City[449] we were informed by them that it was fifty-six miles to the link[450] or sink of this river and we came nine yesterday evening. We are now in a regular desert, nothing but ashes, no grass on the river bottom. The cattle fare but poorly, having to browse on a few willows there along the banks of the river. Met with a person who was from St. Louis. He had worked in the same foundry with old Mr. Bennett, a friend of mine.[451] His son William is dead. We encamped on the top of a bluff bank, not being able to get into the bottom. It was quite a job to carry water up the hill.[452] Very warm and shockingly dusty.

7 September – Travelled about three miles and found grass. Remained there all day. I suppose I have performed my last washing for this trip as I had a great one today, but "oh this darning and mending."[453]

8 September – Started early in search of a spring where we were to take our water for the desert. Found it in about twelve miles. A sulfur spring, and not very cold. Passed over a burning tract of land. It was a kind of turf and sometimes we went up to our knees in ashes and the ground in some places was sunk an acre or more in a piece and then again it was rent asunder as if by an earthquake and fissures ten to twenty feet deep. It was quite a dangerous place to cross it as evidently [it had at] some time been the bottom of a lake. We then went in search of a piece of grass called the Meadow[454] where we [will] take some more grass for the desert; also [to] give the cattle a good feed for the desert is more than seventy miles across.[455] Took the wrong road and were passing the Meadow when we fortunately discovered our mistake, and by turning to the left a few miles brought us to the place. It is a nasty swampy piece of land many miles in length. Hundreds of dead cattle and horses lie in every direction. There are a many wagons here preparing for the desert. Traded ten pounds flour for twenty-two pounds bacon. Cold night. The dust today same as usual.

9 September[456] – The wagons continued to roll in all night which was very annoying. I looked out in the night and such a singular sight I never saw; hundreds of fires, for by this time hundreds of packers and wagons had arrived; it put me in mind of the appearance of the Chinese river I have read about where the fishermen light their torches to catch fish at night. The lights flickering about were like an army of will-o'-the-wisps,[457] and in the morning it presented a curious spectacle: some killing beef, some jerking it, others packing grass up to their middle in the mire. Such a motley set of dirty ragged devils I never saw. Mind I don't mean to insinuate we look any better than the rest. Saw a gentleman from California who has come out to meet his family. The news is good. Gold plenty as ever. Went about four miles down the meadow and cut more grass. Encamped there and we went a mile in the swamp and there we discovered a series of waterfalls; the river spreads all over the swamp. A cow got mired in a slough and in pulling her out her hip joint came out and we had to make beef of her. Cold night.

10 September – Started off for the desert. Took in some water. Passed through a lot of Indians of the Paiute tribe.[458] They have got sickness among them and a doctor was staying a time to prescribe for them. Came to a large lake made by the river. We are now near the sink. Nothing green grows round this lake except a few patches of salt grass.[459] It is a nasty, salty place. Travelled sixteen miles.[460] Very cold day and the wind came from the Sierra Nevada Mountains and smelt of the snow. Spencer started to pack fifty pounds.

11 September – Frosty night. I was on watch. Fed our cattle and they look very bad, the salt having scoured them very much. Came to the sink, a large lake. Here the river sinks into the ground and is no more seen. We nearly upset one of the [wagons], the one that got a dip in [the] Blue River. Two or three of us jumped into the river and we saved it. Started right into a long stretch. In one half day I counted one hundred eleven dead cattle, eighty horses and five mules.[461] Wagons, guns, etc., in great plenty were strewn along the road. Travelled on until supper time and continued on until morning.[462] Dark night and had to walk ahead so that the teamsters could see to find the road. Some had to go ahead and make fires of the wagons to light the teams along. Cold night.

12 September – Stopped a little before day and took breakfast. I was very unwell but still had to walk as the roads were very bad, being deep sand. At daylight we started again, but oh, such a sight presented itself. The road was strewn with dead animals, wagons, cooking stoves, water barrels, harness, guns, clothes, tents abounded, and broken-down cattle left to die. The destruction of property is unparalleled in history except in wartime. I saw fifty wagons in one lot and anyone that wanted [a] fire took one and burnt it up. We cooked breakfast with one. Many people escape without anything at all, glad to save their lives. Many that started with a good team from the States are now left without anything at all, barefoot and nearly naked. We lost three cattle and had to leave a wagon six miles from the end of the desert and return for it.[463] Made [the] Carson River about twelve o'clock and glad we were to get water.[464] It looked like a fair. There were booths, tents, etc., with provisions, liquor, etc., flour forty cents per pound, grass ten cents per small bundle.[465] The Carson River is a nice, clear stream running from the Sierra Nevadas and has timber on its banks, the first I have seen in a long time. There have been many deaths here. One this morning just as we arrived – a Missourian. His brother, poor

fellow, took it to heart very much. It was quite heart-rending to see him I can assure you. There has been a great deal of sickness here by the number of graves there are here: the bad water. I have seen sights on this route that would make the stoutest heart quake. People who venture through here ought to make gold. I would not come through again for all the wealth of California. Nothing grows on the desert to support either animal or vegetable life; no birds except carrion crows and ravens.[466] No grass yet for twenty-five miles. Seven hundred dead animals in forty miles, five hundred wagons. Since yesterday we travelled forty-six miles.[467]

13 September – Moved about six miles and by putting the cattle across the river they got tolerable feed. Saw petrified trees. Remained in camp the remainder of the day. Fixed up a swing for the young ones. The crown of my hat being worn out, I found one with a good crown, cut it out and fixed it in mine. We also repaired shoes, etc. This river is named after a celebrated mountaineer Christopher or, as he is generally termed, Kit Carson.[468] He was the principle guide and right hand man of Colonel Frémont when he explored the Rocky Mountains five or six years ago. Then this route was discovered but not used until last year.[469] We saw him at the upper Platte ferry. He had[470] the ferry and was on his way to California. He is a native of St. Louis and has always been in the service of some of the fur companies which is the reason he is so well acquainted with these mountains. We fixed ropes up in the trees and tied our beef up in the trees to sweeten, as it, being kept close in a wagon, it smelt a little sour. Meat will keep in this climate for two or three weeks good without doing anything to it. Cold frosty nights.

14 September – Started over a stony hilly country and in about twelve miles we expected to strike the river again, but the road was so bad and our oxen weak. We had to leave a wagon and intend to fetch it tomorrow. Made the river in about sixteen miles but no grass. Encamped in some timber and it was quite a relief to hear the wind among the foliage. Cold night; on watch. Made up my mind to pack in the morning as the teams are so slow.[471]

15 September – Hail storm. Cold morning and one of our party went after the wagon and we went on five miles to grass where we remained all day. Sold one of the wagons for $5; cost $100 in the States. Threw away our guns and everything we could to lighten the

loads. Three of the company left and went with some traders. I am preparing for packing.[472] Cold night.

16 September – Went ten miles over a heavy, sandy flat and then struck the river again; good grass. Got my pack ready. Saw a California paper with the death of the President in [it].[473] Heard of it a few days before. Trading stations about every ten or twelve miles. They pack their flour out on mules and have Missourian muleteers with their regular costume of open sided pants with bell bottoms, broad rim high crown hats, singular saddles with wooden stirrups, and such gigantic spurs and dirty shirts, etc.

17 September – After breakfast I started with my pack on my back weighing twenty-six pounds containing a shirt, pair of socks, coat, quilt, pair of woolen mitts,[474] coffee pot, tin cup, spoon, box of matches, ditto[475] pills, a piece of soap and towel, comb, seven pounds of flour baked into little loaves or biscuit, three small pieces of bacon, a pint of ground coffee, and my journal to climb the Sierra Nevada Mountains alone, it being about 175 miles.[476] I expect to get over in seven days. At twelve o'clock I ate my dinner and my shoulders were very tired having come about twelve or fourteen miles. Although my pack is heavy I feel as if I was lightened of a load with being released from Sweasey, the man I came out with. I never had such a thorough contempt for any man before. I travelled over fifty miles and slept under a wagon. The wolves yelled and barked and I was so sore I did not sleep much. Met with a couple of men who pretended to have lost a companion and wanted me to go into the brush and hunt him, but it struck me they wanted my pack as they were but poorly supplied. I went on in haste and lost sight of them. Cold frosty night.

18 September – Got my breakfast and started and overtook an Irishman packing. Undertook to go along with him but he was too slow and I soon left him behind. I then overtook ten more from Missouri and agreed to walk with them. Sold some of my bacon and bread as I found my pack getting heavy and there being trading posts every twelve miles or so I can supply myself again. Snowy mountains ahead. Came to the Carson Valley, one of the most delightful places I have seen on the route. It is walled in by mountains, and on the north they are covered with pine timber of a large growth and such a lot of the most beautiful, sparkling cold water runs in small streams. From now on the top. Arrived at the Mormon Station at two o'clock and took dinner there.[477] While laying down after dinner a snake crawled

under my back. One of the men happened to see it so I got up and killed the intruder. In about four miles we came to another trading post. They persuaded us to take the Georgetown cutoff as it saves between twenty and forty miles. It is a pack route over the mountains, so as some of the party had to bake and had an opportunity to borrow baking apparatus they concluded to stay for the night. Having travelled about twenty-five miles and encamped under a large pine tree, I rolled myself up in a quilt and slept. Cool night.

Nevada

22 September
Georgetown

Lake
Tahoe

19 September

Sacramento River

11-14 August
San Fransisco

27 September 1850
Sacramento
11 August 1852

California

Pacific Ocean

September – Started over the mountains and oh, such a road. Had to clamber up the rocks on my hands and knees and sometimes it was loose sand. Five miles of such travelling brought me to the top of the first ridge and very tired I was. I came to a beautiful large lake of clear water. I have certainly seen the elephant[478] on these mountains. Took dinner on the bank of the lake and there is another range of mountains before me which look down frowningly upon me and seem to say "thou little atom dare you attempt to cross my lofty back." I

57

looked up at the lofty peaks covered with snow and lost amid the clouds and said "well I'm going to try old fellow, and if that naked old gentleman with the scythe and hour glass doesn't cut me down I'll cover it too." With that I fancied the mountain smiled and shrugged up his shoulders in contempt, and on I went. I took the wrong trail and had to cross over a swamp and in crossing I walked into a grizzly bear's den.[479] He had been there very recently. I[480] cleared out quick. Found the trail, but did not find my companions. Ascended one spur of the mountain where on the top the snow was deep but hard, then walked six miles and encamped by a large fire I found made to keep the "varmint" off as these mountains are full of bears, panthers,[481] wolves, etc. Laid myself down with my feet to the fire and a snow bed at my head. Cool, clear, frosty night; moonlight. Ten o'clock. Made thirty-five miles.

20 September – Got an early breakfast for I could not sleep; so cold. After going about five miles I commenced to ascend the third and last great spur of the mountain.[482] After considerable trouble I got to the top in about four miles, gave three cheers, and the rocks echoed again. Passed several small streams, also several mountain torrents where the water came dashing and foaming down the rocks at a fearful rate and made such a roar that it was with difficulty I could hear myself call. Crossed one torrent on a fallen tree and the water dashing underneath many feet below. After walking about four or six miles I came to some people grazing stock. Here I stopped.[483] I made some mush sturrah[484] but not having any milk it made good paste such as cobblers use. That was all my breakfast this morning. After walking about over hills and dales, sandy and stony, I came to another spur of the mountain. I then stopped and took some more paste and ascended the spur walking on until I came to a place where the whole forest was in a blaze. The giant pines [were] from one to two hundred feet in height and [there was a] very large quantity of fallen dead timber. It blazed fearfully and I had in some places to run through the flames and [was] up to the ankles in hot ashes. I got a little scorched. I have seen the oak wood on fire in Illinois; that was nothing to compare with these pine woods. I always thought Cooper's account of the fire in *The Pioneers* was exaggerated, but I [now] believe it true.[485] Such a smell of pitch rosin and turpentine and then the smoke; it was suffocating. Walked until midnight and the moon was of but little service owing to the density of the foliage. Such another bad road I never want to travel.

Walked thirty-five miles and made a fire along a fallen tree and went to rest. Cold night.

21 September – Shower of rain and dull morning. Baked bread in the following manner: mixed the water and flour together on a piece of bark then put the dough round little sticks and put two little forked sticks in the ground and then lay the other across before the fire and baked it so. It rained all day and I started two deer but having nothing to shoot with they made off.[486] It quit raining about midnight. I was very sick, cold, and tired and could not eat any supper. Laid down wet through; also my quilt was wet. Made twenty miles. I have now passed over the worst hills. When I arrived here tonight, I thought of Colman's Mountaineers song in the Opera of the Mountaineers: "Faint and Weary the Way Worn Traveler."[487]

22 September – Had paste for breakfast with a piece of mule I found dead. Came to a log cabin, the first house since I left the States where I found the party I left at the foot of the mountain in the valley. We had been pretty close together but did not know it. Three of the party left our company and went down to the river to mine but the rainy season having commenced they cannot do much until next season, so two besides myself went on. They met with an acquaintance going down to Sacramento city. I think of going with them. Arrived at Georgetown.[488] Such a town; a lot of shanties stuck together in the most rough manner. Lots in tents, others living out of doors. Our road is through the mines. The road and every creek is dug up and everything and body looks strange. I expect to get a job in the city and also I am very anxious to hear from home.

23 September – Started for the city. Ten miles brought us to Coloma, the place where the gold was discovered.[489] Crossed the river at Luther's mill. Lots of people digging. Could have got a job to work on a dam but having to work in the water I was told it would certainly make me sick, so [I] travelled on. It looked very like rain so we encamped under some trees. The rain poured down directly in torrents, burst the gold washers' dam in Coloma, so [we] fetched some slabs we saw a distance off and fixed them over us for the night. It rained all night long. Very wet and uncomfortable. The miners having nothing but brush cabins are in a fix owing to the rainy season setting in sooner than usual.

24 September – Wended on our way. Cool morning and off a hill saw the beautiful valley of the Sacramento spreading out before us.

Passed through numerous small settlements. The road between the mines and the city is all life, wagons passing and repassing all the time, stages, etc. Met a lot of Chinese; a curious looking set, little bits of fellows but appeared to be very merry as they were laughing and talking very much. Went about twenty-five miles and slept under the broad spreading branches of a huge oak.

25 September – Slept but poorly owing to the chamber maid not shaking up the bed well. There were a lot of stones under the blanket. It was a fine night.

26 September – Travelled on until we came to Brighton, six miles from Sacramento. We slept down the bank of the river.

27 September – Started for the city and came to the celebrated Sutter's fort, partly in ruins and used as a hospital.[490] When about a mile from the city we put our coats and packs in an oak tree and went into the city. No work and could not hear of a job at all, there were so many thronging in. Heard they were digging a levee and went to get a job and the first thing I saw was a Mr. Jackson who left us on the Humboldt River. He was digging with a spade and wheelbarrow. We were both very glad to meet. He had no money and I had the sum of fifty cents. Went and fetched my pack, and started housekeeping in the bushes with him. The boss of the works gave me my supper and promised me work next Monday.[491] Have to live until Monday. Went and laid down on some straw.

28 September – Worked half a day for Mr. Jackson as he wanted to go to the post office and I could get my dinner. He got the breakfast and fetched me a piece out in his hand. In the afternoon [I] went all over town to try to get a job. I was very weak. I found I could not stand that kind of work.[492] I went into every place where I saw tools, a hammer or a saw, or shavings. I tried every place but the answer was no. It appeared to me they could not say anything else. At last I found a place where the boss said he would give me work on Monday at five dollars per day. Wages were then ranging from ten to sixteen dollars per day but such a flood of emigrants had made hands too plentiful. I jumped at the job and went and told Jackson. He was glad. We concluded he was to leave his job soon as mine began but he could not leave now because we had nothing to eat. I was taken very bad with the diarrhea; laid in an empty wagon and Mr. Jackson brought out a piece in his hand for me. Fat salt pork and bread is rather tough diet for a sick person but I am getting used to it. We both slept in the wagon.

29 September – Walked about town a little but still sick. Had to lay in the wagon again – very sick.

30 September – Sunday. Mr. Jackson stopped working today. We went about town. There I could see no difference between Sunday and any other day; gambling, saloons and all work was going on as usual. Went up and saw the person who gave me the job and agreed to put the roof on a house for twenty-five dollars, then walked down to the levee where we met with Mr. G. Hamilton who left our train in the early part of our trip. There were now three of us. I spread my job amongst us and with my fifty cents I bought us a loaf and went into the building I was

[Book One of *Journal*] [493]

going to work at in the morning. I went with the boss to see the job and when I returned my loaf was gone. He asked what was the matter; I told him and I suppose he saw by my long face I had no more money. He gave me a dollar and took me to a provision store and I got a ham and some potatoes and I walked on after my comrades who were very much surprised to see me so heavily laden. I also found they had got the bread. We went into the bushes and had a perfect feast. No Queen Victoria ever had such a banquet in the world. We left our provisions up in a bush and strolled about and washed our shirts in the river to go to work in the morning respectable. Felt as well as ever.

[Sacramento, California, probably Fall 1851][494]

My very dear wife,

received your letter dated the 28th April [1851], postmark 1st May. It stated that you had received a draft, but for what amount it said nothing at all about it. This is very wrong; it makes me feel uncomfortable. You speak as if you had[495] saying you did so and so with the ten dollars I sent you. I sent you $200 dollars the last I sent and it was worth mentioning the amount. I have money on hand to send but I don't like to do so for fear you don't get it as you say nothing about it. And another thing, your letter is a regular batch of conundrums. I can't make it out. Now my dear wife, you ought to think that I know nothing of [New] Harmony. You say something about Mr. Bennett's last and then go on to say you know what I mean. How the deuce should I know what you mean? Has Mrs. Bennett run away or something of the sort; be more plain. I should like to have been at home by Christmas but can't. I expect to go to the Sandwich Islands[496] shortly as our company expects to go there. I am now engaged in the same place as when Mr. Jackson left the Pacific Theater.[497] You must tell Mr. Jackson how we have driven Mr. Stark's[498] company out of town. They divided; half to the bay and half to Marysville.[499] They left Marysville to join the other party at the bay and played one night here but they had a very poor house and the audience were guying[500] him all the time, and moreover he is married to the Kirby[501] and such a kick up[502] you never saw. We have had such champagne suppers and such kick ups you never saw. We have crowded houses every night and I am a great favorite of Thorne's.[503] I got a puff in the paper about my scenery. The other day the new theater at the bay burned down and all the company is up here. McCabe[504] was stage manager at the bay and lost all he had. They are fixing up the Lehama,[505] and Thorne is having a new theater built at Nevada [City];[506] also one in this city, next to the Orleans House on Second Street,[507] of brick, and fireproof. We have a boat and go out sailing on the river and bathe every day. Although I am engaged as carpenter I paint and play small parts sometimes. Mr. Stark, I am sorry to say, is fast losing his gentlemanly deportment. His marriage has ruined him. Mr. Crain[508] and Jack Harris[509] both drunk again; a bad job. Both Mr. Mansfield[510] and Mr. Madden[511] are at our

place and Brigham[512] is also engaged, and Billybus Barrybuss[513] sends his best wishes and says when you sprout in the West to remember that imitation is a vile talent. Mr. Fisher[514] is still carrying on his blacksmith shop and Mr. Doyle[515] was in town a few days ago and is gone to Nevada City. Mr. Fisher was up at Nevada [City] last week. He saw several of the [New] Harmony folks. [Of] Mr. and Mrs. Bolton,[516] he says the old man has got a very stingy better half and they are doing well at boarding [and] housekeeping. Mrs. Combs'[517] son and daughter are both well. Bob Robson[518] is gone to the north mines; also Mr. Keal[519] and the Lichtenbergers.[520] Mrs. Pullyblank wants to return, also Mrs. Wilsey and she has a little one and I expect they will return at the beginning of the year. John Williams and lady are also at Nevada [City] and well. Mr. Daniels[521] is still in the tripe[522] business and doing well and has sent for Mrs. Daniels. I believe Ira Lyon is near there, also Oscar Felch[523] and W. Evans[524] and W. Bolton.[525] John Hale[526] is also there. Doctor Cook and lady[527] are also there. Mr. Henry Pratton[528] [and] Doctor Conyngton from Mt. Vernon[529] are there. Sometimes I think of sending for you to come here as I don't think I could like to live in that picayune place, but as you say if you can buy old Twigg's[530] place I will come home, but if Nettleton[531] has furniture it will be a bad chance to do anything there at the business. If you were in this country for three or five years we could get what money we want. The mail closes at twelve and it is now a quarter [to] and I must finish. Mr. Chaffee[532] is living with me yet, but not doing much. He talks of going to the mines in a week or so. Mr. Gullett[533] is at Cold Springs on the north fork of the American River and as to Mr. Wilkinson[534] he has never been heard of but once since Daniels left him. He stopped at a ranch to herd cattle and there is a tale that he was killed along with the rest by the Indians; how true I don't know but I don't believe anything about the hanging of him. John Mills, J. O'Neal,[535] M. O'Neal and F. Duckwork, J. O'Neal's uncle, were in town a short time ago; also Matt Stoker.[536] He had seen Bill Wilsey who was doing very well. I think I have mentioned them all. Now as for myself I am well; was never better in my life. I am terribly bothered about shirts. It costs me fifty cents per day for washing and the washerman has lost me three shirts and sent me three little devils I can't get on. I am about free from lice now. I board at a nice house; one ounce[537] per week. I think Mr. Jackson did very wrong by going home. If he would have stayed until Christmas[538] I should have come with him but I will have a home. Let me know what you have done about Twigg's place, if it can be got, or if you will come out here. With love to yourself and our dear children, believe me to remain your

Affectionate husband

W. F. Pritchard

Remember me to all inquiring friends.

[Book Six of *Journal*]

W. F. Pritchard's Journal Homeward Bound

 August 1852 – Left Sacramento City at two o'clock [in the] afternoon in the steamer *Confidence* with the good wishes of all friends and I must say I have left several warm ones in that city. Several accompanied me down to the steamer and after a pleasant run [we] landed at San Francisco at eleven o'clock [at] night. Put up at the Mercantile Hotel where they charged me one dollar and [a] half for bed and supper.

12 August – Moved my quarters to a cheaper establishment. Walked about all day to see the Lions[539] as I had only been to San Francisco once before and that in the rainy season. At night went to the Jenny Lind Theater to see the great Booth and two sons[540] in Shakespeare's great tragedy of *King Lear* but the old man is not what he once was; but the piece went off very well.

13 August – Walked about, inquired about passage, etc., and found passage very high. Did not purchase a ticket; thought they would be cheaper tomorrow. I look upon San Francisco as a very disagreeable place. The wind blows the sand about and [it] is very annoying to the eyes. Went at night to the American Theater to see Mr. Procter[541] in the drama of the *Wizard of the Wave* [542] and it went off very passably but there was a poor house.

14 August – Purchased a ticket by the steamship *Pacific* for the Nicaragua route and by nine o'clock had my luggage on board and at a quarter to twelve we got under weigh.

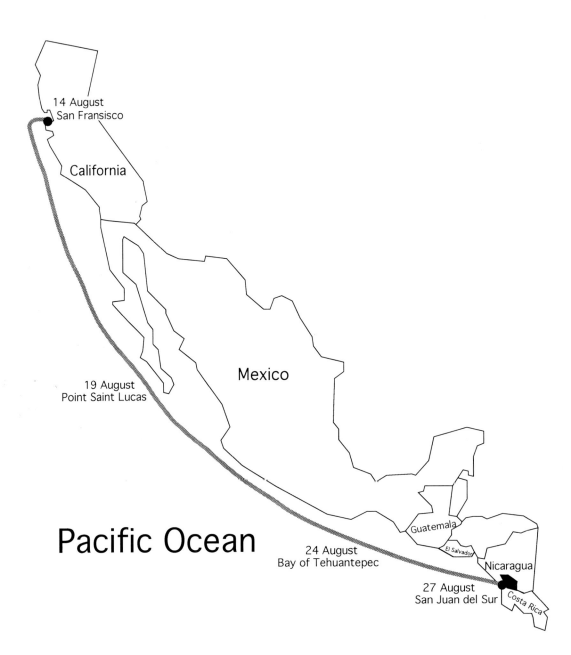

14 August
San Fransisco

California

19 August
Point Saint Lucas

Mexico

Pacific Ocean

24 August
Bay of Tehuantepec

Guatemala

El Salvador

Nicaragua

27 August
San Juan del Sur

Costa Rica

he people on shore gave us three cheers which we
returned. We then fired off a gun and bid farewell to
California. The bay and harbor of San Francisco is one of
the largest in the world and is crowded with vessels from
all parts of the globe. Even the Chinese, a people who
never were known to travel before, are here in plenty and [a] very
quiet, industrious people they are. Passed Goat Island;[543] this island at
the first discovery of the gold was inhabited by numerous herds of

goats but the rifle has destroyed them all, and it is now a desert island. Passed Angel Island[544] – the place where fools retire to settle their quarrels by the code of honor. We then sailed by the old Spanish fort. It is now fast falling into decay. It stands on a point of rock at the entrance of the bay. We then came to the Golden Gate or inlet from the Pacific Ocean. This name was given to this place by Colonel Frémont, there being a hole or natural archway through the rocks that a small craft might pass through. Passed the Farallons,[545] a group of small islands inhabited by sea fowl. Some people make a living by making trips to these islands for eggs which find a ready market in both [the San] Francisco and Sacramento markets. They are large and of a bluish cast with brown and black spots but taste a little strong; but where eggs are scarce as they are in California at present, they are a great luxury and from twenty-five to fifty cents each. Very cold and rough. The berths being all taken up, I had to sleep on deck. Rolled up in a couple of blankets. Took a drink of brandy and lay down. A good many seasick.

15 August – Got up early; a very cold morning. Did not sleep much, not having got used to the motion of the vessel yet. Took on [a] breakfast of bad coffee and worse biscuit and meat. The weather [was] foggy and cloudy; nothing [was to be] seen but the broad Pacific Ocean and an occasional albatross flying about except the scenes on board.[546] There are a good many western people on board who never had a sight of salt water before, and indeed some of them were laughable enough. One in particular who cried out "New York," an expression a person is sure to say when he vomits or at any rate it sounds very much like it, and between every retch he cried out "If I was only back I would sooner cross the plains forty thousand times." Once the vessel gave a lurch and one poor sick fellow was jammed into a side of beef which [had been] hung up close by. "Oh! damn California; if I was only home" cried he and looked pitiful, but a parson would have laughed to have seen him. I have felt no symptoms of sickness yet. Had an exhortation by a smooth faced Methodist and the promise of another this afternoon. Had rice and molasses for dinner. Went early to bed. Not so cold or so rough.

16 August – Slept but little and was awakened by the mate's cry of "Wash decks." A wash in salt water and then breakfast. We had some meal mush and molasses. A beautiful and warm day. Some of the pas-

sengers amused themselves with shooting at the albatrosses, others [by] playing cards, reading, etc. Went to bed – calm night.

17 August – Got up after a good night's sleep. Beautiful morning. The Pacific this morning was truly the Pacific. Nearly calm, so our trust is now in steam altogether. Saw quite a number of porpoises.[547] I think they were much larger than any I saw before. They came pitching quite close to the vessel. The western folks on board stared and said they look like hogs. In the evening [we] passed an American ship bound for San Francisco. We exchanged signals. It is a beautiful sight to see a full rigged ship at sea with studding sails low and aloft, although steam is the thing to go ahead and that's what we want in our days.

18 August – I was awakened by the cry of a whale.[548] I got up and the sea was literally alive with the monsters spouting and topping about. They looked like the bottoms of large ships and sometimes they would throw their huge bodies quite out of [the] water, and to kill the monotony of the season we had a fight between the cabin boy and the baker. Saw a quantity of flying fish.[549] Bad food today again but won't I make the eggs and chickens fly when I get home. Indeed the food is enough to give us the twist in the bowels for I cannot eat enough to keep them filled. Took the vote of the passengers to see how the candidates for the President stood. Scott[550] [had a] fifty majority.

19 August – Shortly after breakfast I was startled by the cry of "Land!" There it was sure enough on our larboard bow.[551] Point Saint Lucas – the point of land which forms the Gulf of California.[552]

20 August – Being now in the Gulf of California we see no land. A few whale and flying fish are all that are to be seen. Very very hot – awful.

21 August – Warm night but little sleep. Last night [there was] a fine rain, the first I have seen for several months. "Land ho!"[553] We are across the gulf and saw the other point although no land [is] in sight now. "Sail ho!" the cry arose at dinner. Down went knives and forks and a general rush took place to see [it], but the vessel was so far in the distance we could not make out what it was. After nearing it we discovered it was a brig.[554] As we did not pass near enough to exchange signals we could not tell what country it belonged to or where [it was] bound. Shortly we saw the remains of an old boat. Killed an old lean steer; the butcher struck him back of the horns and he fell. That is the way the Mexican bull fighters kill the bulls.

22 August – A shower of rain and considerable head wind. "Land" – a small mountain was seen for about an hour.[555] An ox died and was thrown overboard. The preacher gave us a morning and evening exhortation and if I was him I would put my head in a sack, for of all the botch preachers I ever heard he beats all. A perfect hypocrite. Rough night.

23 August – The sea [is] calmer this morning but the vessel rolls awfully. Bad coal from Sydney. "Sail ho!" Fine night.

24 August – Rough morning in the Bay of Tehuantepec;[556] getting short of water. Came very near having a fire; the mattress caught from the stove pipe. Soon got order by throwing them overboard. Fine night but the vessel rolls badly.

25 August – Fine view of volcanic mountains.[557] They are on fire, but being daylight the fire cannot very plainly be seen. These mountains were in sight all day and at times it looked grand to see the tops of them above the clouds. Smooth sea and the vessel [was] making good time. In the distance it looks like rain. Bird roosted on the foreyard.

26 August – Thursday – Just as we got to bed, or rather to lie down, there came on a most awful storm. It rained and blew with lightning and thunder. Truly awful and the vessel being light owing to us having got near our destination. It lasted all night and a truly miserable night it was. We were all wet through and lay down wrapped up in wet blankets. The sea frequently makes clean sweeps over us. Sick men groaning, some praying, others cursing, and so it was. Lots of porpoises and black fish.[558] A person, a friend of the captain, got up a paper for us to sign that we were well treated, etc., but as it is rather a difficult thing to persuade people they have been well treated when they have not, there was a counter set of resolutions got up and denounced the others, also the officers, pretty strongly. About eight o'clock the cry of "Land ho!" put everyone on the lookout. There it was sure enough but we had to lie to until morning. A brigantine[559] passed under our stern and [we] hailed [them] as [to] how far off San Juan [was].[560] The reply – "About ten miles."

27 August – By daylight we were under way and ran into the romantic and snug little harbor of San Juan del Sur.

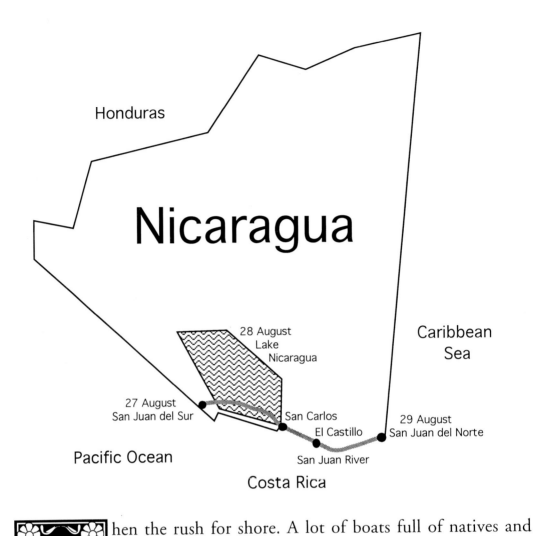

Honduras

Nicaragua

Caribbean
Sea

28 August
Lake
Nicaragua

27 August
San Juan del Sur

San Carlos

El Castillo

29 August
San Juan del Norte

Pacific Ocean

San Juan River

Costa Rica

hen the rush for shore. A lot of boats full of natives and Irish (I wonder where they are not). A considerable sea running made it very bad to get into the boats but after a deal of swearing and some fighting several got off. I got in with my things and made for shore. Here again was another stoppage. There was so much sea running which caused a great surf on the beach. A lot of natives naked, in fact they are all naked or nearly so, waded into the water and seized our luggage and packed it to shore. In the squabble to obtain a job they ran the boat too high on the beach and two seas washed over us wetting my back. I then got on the back of a native and was packed ashore with a hat box in my hand. The next thing [was] to pay the natives and then get our things packed up to the company's office where they were weighed and ticketed. We then got a ticket for a mule each. It then began to rain and when it does rain here it doesn't make a fool of it. No breakfast – all eaten up. Mounted my charger – such a rip[561] – and off we started. I carried a rifle over my shoulder with a hat box on it. I was wet through directly. We travelled twelve miles this way. They are making a road through and there are a great number of natives, also Irish, at work on it and when finished I have no doubt it will be a good road, but at present it is awfully hilly, rocky, and muddy.[562] My mule got mired down and I got some natives to dig it out. When I got through you may guess the beauty I was. My two hats were spoiled after all the trouble I had taken with them. I got some dinner and looked through town.[563] A many did not get through as soon as I did. Some left their mules in the mire, others gave out and were left. One man got off his mule to get some wild fruit and the mule ran away and a preacher and myself tried to catch him but in doing so I ran into a hornets[564] nest and was badly stung. This little town has been built since the route was opened and is like all the other towns in Central America: built with bamboo poles and covered with plaster and leaves. The boat being ready we went on board and lay there all night. Lake Nicaragua[565] is a beautiful sheet of water but full of alligators[566] and sharks, etc.[567]

28 August – Started early for the San Juan River. The scenery on and around this lake is very romantic and singular. The lake is literally full of islands, so few inhabited by natives, and one has a town on it. We travelled thirty miles on the lake, then arrived at the fort and town of San Carlos.[568] This being a port of entry we had the customs house officers on board a few minutes and then started down the river thirty-eight miles. We then changed boats and proceeded on to the Castello Rapids.[569] Here we changed boats again. This is an old fortification of the Spaniard – an old castle stands on a hill but now the guns are dismounted and the fort in ruins is overgrown with vines, etc. The boats we take here are all stern wheel boats and small on account of the shallowness of the river. We have to travel ninety miles or so. We had a very narrow escape of being stove in. They have a dredging boat clearing out a channel and it happened in one of the worse places in the river – in a rapid. We ran along side of it and struck it without any further damage than the loss of our yawl.[570] It was a beautiful moonlit night with an occasional shower. Before night some of the passengers amused themselves by shooting at the alligators as they lay basking on shore or on an old log. We arrived in the night at

71

the company's yard and took in coal, then crossed over to the town of San Juan del Norte.[571] By this time it was morning.

29 August – Went to the Vivian Hotel kept by an American lady. This town comprises one long street and a square. The principal buildings are hotels and the rest are native shanties. Got a good wash and some breakfast. Bought a green parrot[572] and hired a boat which took us on board the steamship *Daniel Webster* bound for New Orleans.

3 September
New Orleans, Louisiana

Gulf
of
Mexico

29 August
San Juan del Norte, Nicaragua

Fired a gun and started by seven o'clock [in the] evening. Went to bed and fell asleep.

30 August – Monday – Nothing of note, but a much better regulated boat than [that on] the Pacific. Clean with the best of food and gentlemanly officers. Considerable showers.

31 August – Showering and going at a rapid pace. Felt bilious and took some medicine so that I was very unwell. Parted with a good deal of bile. Got some tea and went to bed.

1 September – Got up much refreshed. Took a glass of soda water and then, after a while, breakfast. Although [we have] a smooth sea we are not making much progress. A heavy shower. The boiler [is] out of order. Had to stop about two hours and fire it. Started again and ran about three hours when we had to stop again, the boiler being cracked. It was patched up and we got under way but [we] have lost nearly half a day and we are going very slowly. A poor little bird, a kind of snipe, who had got driven off from land, lighted on the vessel and was captured.[573] The boiler I hope will stand it for they are cracking on steam. Pretty freely showering.

2 September – Nothing of note except that we had plumb dough[574] for dinner and the sailors are holystoning the decks; that is they go down on their knees and scrub with a piece of sandstone and sand. Others tie a cord round a larger stone and pull it backwards. It makes the deck very clean. Good night – wind blowing.

73

3 September – All got up in good spirits as we expect to see the mouth of the Mississippi.[575] Shower. Made the mouth at night and took a pilot on board and commenced to ascend the river. Got over the bar and then took a river pilot who will take us up to the city. Beautiful night.

4 September – Fine morning. After a while it threatened rain but kept off. Passed English Turn where the English fleet was obliged to anchor shortly before the battle of New Orleans. The English landed here and were marching under Packenham[576] towards the city when they were met by Old Hickory, as he was called, General Jackson,[577] and the brave American volunteers who gave the British just what they wanted; a good beating and may they always get the same when they try to trample on the people's liberty. I saw the tree, a live-oak [or] an evergreen, where Packenham fell and died. Got to the city at [578] o'clock, so making the journey from San Francisco in less than twenty-one days. The shortest trip ever made – a distance of five thousand miles.

Hunted round and found a few of my old acquaintances where I lived there,[579] but the majority were gone to their long homes.[580] Took the steamboat *Midas* for Cincinnati. I got off at Mount Vernon, Indiana and then I had only fourteen miles to home and the nearer home I get the more impatient I become. The idea of meeting my boys in the street and not to know them – for two and a half years in little ones makes a great difference and then there's one I have never seen.[581]

NOTES

[1]In the journal he says in the entry for 24 June that he was 32 years old that day; family tradition has 13 June as his birthdate – see the notes with the main text in the entry for 24 June for more on his birthdate.

[2]His occupation is described this way in the baptismal record for William Fowler Pritchard in the Registers of Baptisms, Burials, and Marriages˙ for Ellesmere. It is similarly reported in the baptism record of W. F. Pritchard's sister, Juliana Mary Pritchard which occurred 9 September 1821 in Hordley, near Ellesmere (Shropshire County Records Office, 3945/Rg/3).

[3]In British usage, attorneys are divided into solicitors, who are not members of the bar and who appear only in lower court, and barristers, who are member of the bar and may appear in higher courts (William Morris, ed., *The American Heritage Dictionary of the English Language* (Boston: American Heritage Publishing Co., Inc. and Houghton Mifflin Company, 1969), s.v. "solicitor,").

That he was a solicitor is attested on his tombstone in the yard of the Church of the Blessed Virgin Mary, Ellesmere, where he is styled "Solicitor of this Town;" *Pigot's & Co.'s Directory* for 1842 and the *Post Office Directory of Shropshire* for 1856 specify that he was "solicitor & clerk to the commissioners of land, assessed & property taxes" (p. 51).

[4]This information is from the *International Genealogical Index* (1984); the *IGI* also contains a reference to a Juliana Fowler from Hordley, a small town near Ellesmere, but the original records were not found by Phil Pritchard when he was doing research in the Shropshire Country Records Office in late 1994. Given that the father of the Juliana Fowler from Staffordshire was named William like her son, it is likely that she is the mother of William Fowler Pritchard.

[5]Considerably more is known of her mother Dorothy's family; Elias Needham, her great-great grandfather, was an attorney in Chesterfield, Derbyshire, England where the entry in the Schedule of Graves in Churchyard (1933) says that "his character is best told by the poor." Other information on her family can be found in the *IGI*, the Pritchard family Queen Anne Prayer Book (which came from the Needham family), and the Derbyshire County Records Office.

[6]Thomas Pritchard had moved to Liverpool some time before his son William left for the United States; there are several references in the journal which would suggest that he had lived Liverpool quite apart from the several records from Liverpool relating to the family, including a possible remarriage of Thomas Pritchard.

[7]National Archives, Ship Arrival Lists, New Orleans, Roll 22 (2 November 1841-31 August 1843).

[8]Robert Owen spent parts of the years 1824-1827 in New Harmony (Anne Taylor, *Visions of Harmony: A Study in Nineteenth Century Millenarianism* (Oxford: Clarendon Press, 1987), 81-161).

[9]William E. Wilson, *The Angel and the Serpent: The Story of New Harmony* (Bloomington: Indiana University Press, 1964), 5-29 on the Rappites, 80-81 on

Owen, 199 on the Geological Survey, and 202 on the theater company. On W. F. Pritchard's involvement in the theater, as well as that of his fellow-travelers James P. Bennett and Jonathan Jackson, see "Excerpts from New Harmony Newspapers...," *Newsletter of the Descendants of William Fowler Pritchard* 2 (April 1988): 18.

Several of the people on the trip to California were also interested in the theater, specifically Jonathan Jackson (see his note in the entry for 29 August 1850) and William Bolton (see his note in the letter home).

[10]There is a photograph of this house in 1988 and the location of the lot is shown on a map in Phil Pritchard, "The Properties of William Fowler Pritchard in New Harmony, Indiana," *Newsletter of the Descendants of William Fowler Pritchard* 3 (December 1988): 4-7. Given the dates of the sources upon which this article is based, it is clear that the family did not own this house before William Fowler Pritchard's trip to California, but it is not known whether this house was the first owned by the family, and whether this house was the "Twigg's place" referred to in the letter home. There is an indenture for the purchase of this house by William Fowler Pritchard dated 24 October 1854, so it is likely that if it is not Twigg's place, it is the house the family bought with his earnings from California (photocopy of the indenture from Martha Leigh as reported in Phil Pritchard, "More on W. F. Pritchard's House in New Harmony," *Newsletter of the Descendants of William Fowler Pritchard* 8 (May 1993): 18.

[11]See the final note for 1 July for a possible reason he was tied to Sweasey.

[12]There were few African-American emigrants (George R. Stewart, *The California Trail: An Epic with Many Heroes* (Lincoln: University of Nebraska Press, 1962), 122); William Fowler Pritchard on 26 June says that a party of sixty-five men had two Negro women along to cook for them and on 13 July mentions a Negro woman singing hymns.

[13]*The New Harmony Advertiser*, 20 August 1859, contains a complaint from William Fowler Pritchard, undertaker, for the failure of one of his customers to pay his bill.

[14]Phil Pritchard, "William Shakespeare Pritchard," *Newsletter of the Descendants of William Fowler Pritchard* 4 (May 1989): 3.

[15] The New York publication date is questionable because while this date appears on the title page, that title page also has New Harmony as the place of publication and Times Print as (probably) the publisher – Edward Eberstadt's role is mentioned on the inside of the cover, without a date, but in a way which suggests that it was printed somewhat later.

[16]There are several other sources which refer to individuals named James Bennett:

(1) There is a card for a James Bennett in the Local History Card file of the New Harmony Workingmen's Institute which describes him as the son of a Welsh fur trader and an Indian woman from northern New York and gives his dates as 1813-1869. This would have made this James Bennett 37 at the time of this trip. He lived in New Harmony from "an early age" and married Maria Pooley there in

1835. He went to Kentucky in 1830 but was back in New Harmony by 1846, when he began publishing the *Western Atlas*. When that became unprofitable, he founded and published the *Gleaner* in 1848 and 1849. He was a hunter, as was the James Bennett on the trip.

(2) The editor of the *New Harmony Times* which first published Bennett's account of the trip says that the Bennett who was on this trip was "a member of Robert Owen's community" who attended the School of Industry and founded the *Western Atlas* (later the *Gleaner)* and that he was 56 at the time of his death in 1869.

(3) There was a James P. Bennett who was William Fowler Pritchard's friend who came to New Harmony with him from Covington, Kentucky in 1847 and who worked with him in the theater in New Harmony. Interestingly, Jonathan Jackson, one of the men on the trip, is reported to have come with these two from Covington, and was also involved in the theater; a James Warren is also listed as having come at this time and participated in the theater in New Harmony, but it appears he did not go to California with William Fowler Pritchard and Jonathan Jackson. This data is, I believe, from a note in a New Harmony newspaper and was reprinted in the *Newsletter of the Descendants of William Fowler Pritchard* 2 (April 1988): 18, which did not specify the source.

William Fowler Pritchard refers to this last James Bennett as "old Mr. Bennett" in the entry for 6 September 1850 and makes it clear that he is still in New Harmony. It would thus appear that the first two sources refer to the James Bennett who was on the trip and that we have no further information on William Fowler Pritchard's friend, especially on how much older than William Fowler Pritchard he was to rate the appellation "old."

[17]The spellings of some of these names are uncertain — see the notes in the journal itself where the individuals concerned are first encountered for a discussion of the proper form of their names.

[18]The count at this point includes 2 wives and at least 2 children: the Sweaseys' boy and girl. William Bolton was the Boltons' son but his age is unknown; he is unlikely to have been a minor as his father was 52 at the time of the trip and his father's corn meal mill in 1849 was named "Bolton & Son" (see also the note for Samuel Bolton to Bennett's description of 20 May). William Faulkner is described by William Fowler Pritchard as a "fine young man" at the time of his death on 9 July, but his actual age is unknown; since Bennett calls him "Mr." Faulkner on 7 July and he was apparently travelling without his parents, it is unlikely that he was a minor. Bennett, 2, mentions the number of wagons but not the number of people who started from New Harmony.

[19]A John Corbin, Sr. and wife are mentioned on a special card in the Workingmen's Institute Local History Card file, the "California Argonauts 1850" card. A number of their relatives may also be mentioned, but the way the card is phrased makes the surname of these people uncertain.

[20]Earl H. Pritchard's handwritten notes on his copy of the California Argonauts 1850 card indicates that all the Craddocks were in a group of 1852 emigrants.

[21]Stewart, 297. William Fowler Pritchard does not mention any bridges although he mentions numerous ferries.

[22]Stewart, 292, 296, on the number of emigrants including the proportion of women and children, 135 and 297 on the 1850 route, 293 and 301 on abandoned wagons, and 296 and 319 on the Panama route. Stewart does not mention Nicaragua in his remarks on Panama in the section for 1850; but in later remarks he says that both routes were in use from 1849.

[23]W. F. Pritchard expressed this date as "1st of April;" this is British practice. Most of the dates are just "2nd" and the like, the month being given only for its first day. In contrast, throughout this edition the style specified in *The Chicago Manual of Style*, 13th ed. (Chicago: The University of Chicago Press, 1982) has been adopted.

James Bennett, another member of the party, says that 1 April 1850 was a Monday (James Bennett, *Overland Journey to California: Journal of James Bennett Whose Party Left New Harmony in 1850 and Crossed the Plains and Mountains until the Golden West was Reached* (New York: Edward Eberstadt, 1906), 2.

[24]The journal very often uses the version of "&" that looks like a "+" sign with a loop for "and" and uses an extension of this sign with a following connected "c" for "etc." These have all been spelled out in full in this edition.

[25]Bennett, 2, says that there were eight wagons. Stewart, 115, says that three yoke of oxen (i.e., 6 oxen) was the most usual team for the typical emigrant wagon.

[26]New Harmony is located some twenty-four miles northwest of Evansville in northwestern Posey County which is the southernmost and westernmost county in Indiana. See the Introduction for further information on this town.

[27]Considerable additional information on this person and this work may be found in the Introduction.

[28]The Local History Card file says that Miles Edmonds' wife was Annie Allison (1840-1930) and that he married her in 1859.

[29]Of these places, only Grayville (on the Wabash some 10 miles north-northwest of New Harmony) and Albion (the county seat of Edwards county and another 10 miles in the same direction) show on the *Illinois Highway Map, 1985-86.*

All geographical names in the notes were checked against both the *National Geographic Atlas of the World,* 6th ed. (Washington, D.C.: The National Geographic Society, 1990) and the appropriate state highway map, in that order. If a name could not be found in either of these sources, *in some cases* it was further checked against one or more of the topographical maps prepared by the U. S. Geological Survey. If a name could not be found, this is referenced to the *last-checked* source.

[30]In the entry for 10 September, William Fowler Pritchard reports a Spencer starting to pack (i.e., leaving the company on foot). There is a Thomas Spencer mentioned in the Local History Card file; he is reported as publishing (in 1827) a

notice to the effect that his wife has left him and that he repudiates any debts she may incur.

[31]Bennett, 3, spelled this "Carlisle." Of these places, only Carlyle shows on the *Illinois Highway Map, 1985-86.*

[32]Now the Kaskaskia River, which has a dam just above Carlyle, producing Carlyle Lake.

[33]W. J. Sweezy had a meat shop in New Harmony, 1848-1849 (Local History Card file). The Foreword to Bennett, 1, has him as "Sweesey" while virtually all other references to him have "Sweasy" (there is one to "Sweazy" (22 July)). His son, Richard, is called "Sweasy" on p. 36. William Fowler Pritchard spells his name consistently "Sweasey" (21 July, 10 August, and 17 September) and I have chosen to use this version. He is listed as "Sweezy" on the California Argonauts 1850 card.

[34]This was probably *not* Ammon Lyon, born ca. 1820's, gunsmith and engineer, father of Lution (1849-1898), who died on the plains according to the Local History Card file. He was the partner of A. E. Fretageot, Sr. in a saw mill in New Harmony from 1836 onwards. He had a brother, James Lyon.

An Ira Lyon is mentioned in the Foreword to Bennett and by William Fowler Pritchard in his letter to his wife, below. Bennett almost always uses "Lyon," in one case with "Ira," but uses "Lyons" once, on 2 April.

[35]John Mills (Bennett, 1-2). William Fowler Pritchard first mentions Mills (without a first name) on 1 July. The Local History Card file has a card for a John Brackett Mills, born in 1822 and still living in 1922. He lived in New Harmony until 1835, then moved to Evansville. As Evansville is only 15 miles from New Harmony, it is possible that he joined the residents of his former town to make the trip to California.

[36]Or Combe (unclear in the manuscript, in his letter home) which is also the form that Bennett uses, everywhere (4 places) except 10 April), or Coombe (Bennett, 10 April (2 times)). Possibly the son of Mrs. Melissa Combs, who was the daughter of Mrs. Killian Lichtenberger (1787-1855) according to the Local History Card file. This person is not mentioned in the Foreword to Bennett as coming from New Harmony, nor is he mentioned as having joined the party later as Spencer is (see below). William Fowler Pritchard's mention is of a Mrs. Combs' son and daughter; since it is made in Sacramento, we can't know for sure if she came on the trip or went by sea and whether either of her children was born when the trip was made. Bennett, 22, also mentions a *Miss* Combe. There is a Mr. Coombs in the California Argonauts 1850 card in the Local History Card file.

[37]See Stewart, 113ff., on the different types of draught animals used; he does not, however, mention the speeds of the various possibilities. William Fowler Pritchard explains how they calculated their speed of travel in the entry for 31 May, dividing by which, they must normally have been on the move for 5-9 hours a day.

[38]George Hamilton, who played the violin (Bennett, Foreword and 17).

[39]This is the current spelling of the name; Bennett, 5, spelled it Cain.

⁴⁰Mitchell is not mentioned in the Foreword to Bennett as having come from New Harmony but is also not mentioned in the text of Bennett as having joined at some later point. He is first mentioned in the entry for 15 April (Bennett, 5). There are two Mitchells mentioned in the Local History Card file: Dr. Elisha Vansant Mitchell, married to Amanda Cox [a Thomas Cox is listed on the California Argonauts 1850 card] who died in 1865 in New Harmony, and a Dr. Samuel Minton Mitchell who married Mrs. Martha Jane Peas of Blairsville in June of 1845 (from the *Indiana Statesman*).

⁴¹A tributary of the Mississippi which runs between that river and the Illinois River.

⁴²Called Spencersburg by Bennett, with a recent population of eight *(Missouri Official Highway Map, 1987-88)*.

⁴³This town does not appear on the *Missouri Official Highway Map, 1987-88.*

⁴⁴A tributary of the Mississippi which had several forks at the time, two of which the party would probably have had to cross. Currently, Clarence Cannon Dam is downstream from the locations they would have used and they are now buried under the resulting Mark Twain Lake *(Missouri Official Highway Map, 1987-88)*.

⁴⁵This town does not appear on the *Missouri Official Highway Map, 1987-88.*

⁴⁶Bennett, 8, described this town as "miserable, dilapidated looking."

⁴⁷Bennett, 9, on 6 May says that on 30 April they entered the trail followed by some Mormons a few years earlier on their westward trek across Missouri. This is incorrect; in 1847 the Mormons started from Council Bluffs, Iowa, and never had any reason to enter Missouri, whence they had been driven several years previously *(Encyclopædia Britannica, Micropædia, 15th ed., s.v. "Oregon Trail")*.

⁴⁸This town does not appear on the *Missouri Official Highway Map, 1987-88.*

⁴⁹A southward flowing tributary of the Missouri, which is some 5-10 miles south of where they made this crossing.

⁵⁰An eastward running tributary of the Grand River which runs just north of Kingston, which is 44 miles northeast of Kansas City.

⁵¹Stewart, 112, says that most wagons used by emigrants did not have springs or brakes.

⁵²Reported as married in William Fowler Pritchard's letter to his wife from Sacramento, below. William Fowler Pritchard consistently spells his name "Pulabank." Mrs. Josephine M. Elliott of the Workingmen's Institute in New Harmony remarked in a letter to Earl H. Pritchard that the Pulleybank or Pully-blank family is "encountered often in the life of the community," although no Local History Card file references were given.

A grandson of William Fowler Pritchard (Thomas MacDonagh Pritchard) knew of a Fred Pulabank who lived in Ellery, Illinois and whose father, Jack, came from New Harmony (recollections of Earl H. Pritchard).

⁵³Stewart, 119, says that ca. 200 lbs. of flour per adult was the normal supply.

⁵⁴Not to be confused with the Platte River which they followed later, which is a major tributary of the Missouri River running through Colorado, Wyoming,

and Nebraska. The Platte referred to here flows to the north of the Missouri and joins it just north of Kansas City; the larger Platte River joins the Missouri just south of Omaha.

[55]Bennett, 11, only says that these cows were "missing," not specifically "stolen" as William Fowler Pritchard's first entry, for 19 May, below, states. Bennett's entry for 21 May (p. 11) says of what is probably another incident (which occurred on 20 May) that seven cows were initially missing but that five of them were found.

[56]Born in Bolton, near Manchester, England, in 1798, died in New Harmony on 24 February 1870. A wool carder and charter member of the Workingmen's Institute (1838). He arrived in New Harmony in 1826. He was married to Elizabeth "Emma" Hodgson and had a son, William. Emma died on the journey on 21 July 1850, aged forty-seven. He had a corn meal mill in New Harmony in 1849 which was named "Bolton & Son," suggesting that William was no longer a minor at this point. He eventually returned to New Harmony where he died (memoirs of another son, Frank DeVoltaire Bolton mentioned on Bolton's Local History Card file card). There are no publications by Frank Bolton listed in either the *Dictionary Catalog of the Research Libraries of the New York Public Library 1911-1971* (New York: The New York Public Library, 1979) or *The National Union Catalog Pre-1956 Imprints* (London: Mansell, 1972).

[57]He was Canadian born (Bennett, 31). William Fowler Pritchard mentions him in the entry for 29 August. There are two Moores mentioned in the Local History Card file: Mrs. Amanda Moor (1842-29 December 1906), and Mrs. William (Arvilla) Moore (1821-12 March 1924) who was born near Mt. Vernon and whose husband died ca. 1906.

[58]The Local History Card file mentions a Zachariah Wade who came to New Harmony from North Carolina in 1807; the Wade on this trip may have been his son Isaac Murphy Wade (born in 1827). Mrs. Elliott mentions that John Wade is the only Wade known to have gone to California, but no reference for this information is given.

[59]Nothing is known of this man. The Foreword to Bennett, 1, names him "Fever," but Bennett's text itself always uses "Fewer" (6 and 22 August).

[60]Stewart, 127, notes that *early* May was the usual date for *departure* from the starting point. The original (1841) starting point was Independence, south of St. Joseph. Council Bluffs, north of St. Joseph, was added in 1844, St. Joseph itself was added in 1845, and Old Ft. Kearny, just south of Council Bluffs, was added in 1846 (p. 131).

[61]There is a row of cliffs about 200 feet high five miles to the west of St. Joseph. St. Joseph is on the eastern bank of the river and the river bottom separates it from the bluffs. Since they run directly north-south, a six mile trip from St. Joseph would have put the train a mile or two north or south of St. Joseph. Possibly they were at what would then have been a bulge in the river a bit north of St. Joseph which at that time ran almost to the cliffs and which today is Browning Lake and cut off from the river. No part of this row of cliffs is actually labeled "The

Bluffs" (U.S. Department of the Interior, Geological Survey, 1:24,000 quadrangle for Wathena, Kansas).

[62]In modern day Kansas, this area is part of the Sac and Fox Indian Reservation and the Iowa of Kansas Indian Reservation; the Kickapoo Reservation is some 20 miles further west (*Kansas Official Transportation Map*, 1984).

[63]There are two genera of rattlesnakes and many species; these were probably Prairie Rattlesnakes, *Crotalus viridis viridis* (Raymond L. Ditmars, *A Field Book of North American Snakes* (New York: Doubleday, Doran & Co., Inc., 1949 {overwritten 1945 on the copy in the New York Public Library Research Library}), 244, 251).

[64]The text of book two says "20 May – Crossed over the Missouri River into the Indian territory. One man killed twenty-four rattlesnakes."

[65]The text of book two says "Travelled about six miles. Saw several Indians of the Kickapoo Tribe."

Bennett, 11, says they didn't travel at all, the day being spent in search of lost cattle.

[66]The text of book two says "Travelled about sixteen miles."

[67]The text of book two says "Travelled eighteen miles. Heard of the cholera being ahead. Saw several graves of people who died on the trip."

Bennett, 12, says they passed the Indian Agency about eleven a.m. He also says that this is where the "American Plains" start and that they paid the Indians tribute.

[68]Bennett, 12, says they went 19 miles, passed Bear Creek and saw returning emigrants.

[69]Albert Dexter came to New Harmony with his brother Simeon (1811-39) and sister-in-law in 1837 from Royalston, Massachusetts (Local History Card file). He is named on the California Argonauts 1850 card.

Bennett, 12, indicates that this is when Dexter joined the party.

[70]William and Lucretia Stoker Wilsey. "Bill" Wilsey is mentioned by name in William Fowler Pritchard's letter to his wife, below. Bennett, 12, specifies that they caught up with a Mr. Corbin of Dexter's party, but he does not mention the Wilseys.

The letter preceding "Wilseys" is unclear in the manuscript: it looks just like the "I" which begins the following sentence but it should be a number greater than one because "Wilseys" is plural. The California Argonauts 1850 card mentions them, and specifies his wife's full name (Lucretia Stoker Wilsey); it is possible that she is related to the "Matt Stoker" mentioned in William Fowler Pritchard's letter to his wife. Both husband and wife are mentioned in the letter home as well. Thus, the best guess is that "2" is intended. There are many Wilseys mentioned in the Local History Card file according to Mrs. Josephine M. Elliott of the Workingmen's Institute.

[71]"Nin-a-haw" in the text. Either the Big or Little Nemaha. The Little runs north of the Big and its northwestern part is on one of the known trails although

no mention is made of crossing the Big Nemaha or Wolf Rivers which intervene. Bennett, 12, calls it "Nemehaw Creek."

[72]One of several varieties of sucker found in the Mississippi Valley, either *Ictiobus cyprinellus* (Bigmouth Buffalofish), *Ictiobus niger* (Black Buffalofish), or *Ictiobus bubalus* (Smallmouth Buffalofish), these fish are deep bodied, heavy-headed, and prolific (Henry Hill Collins, Jr., *Complete Field Guide to American Wildlife* (New York: Harper and Row, 1959), 531). The sucker family is also mentioned in the entry for 2 August.

[73]*Esox lucius*, or Northern Pike, a pike-snouted fish which can weigh up to 46 pounds (Collins, 529).

[74]There are no obvious narrows, and no features named such on the U.S. Department of the Interior, Geological Survey, 1:100,000 30x60' map for Blue Rapids, Kansas.

[75]Bennett, 13, says 13 miles and that they were now in Pawnee country. William Fowler Pritchard mentions that they were in Pawnee country on 2 June (i.e., 6 days later), and they would still have been in their territory, which was most of Nebraska.

[76]W. F. Pritchard spells it "caral." Bennett, 16, spells it "caralie."

[77]An extensive southward flowing river with several branches, tributary of the Kansas River which is a tributary of the Mississippi River. It begins within a mile or two of the Platte River (i.e., the Platte has no northward flowing tributaries in central Nebraska – the area all drains south into Kansas).

[78]Of the willow family (Salicaceae) and one of the poplar genus *(Populus)*, probably the great plains cottonwood, which is characteristic of stream beds in the plains. A rapidly growing tree with greenish-gray bark and very soft wood (which is thus used only for pulp, not lumber). Well adapted to the wind-swept prairie (Clarence J. Hylander, *The World of Plant Life* (New York: Macmillan, 1939), 175-78). The great plains cottonwood is *Populus sargentii (Webster's New International Dictionary of the English Language,* unabridged ed. (Springfield, Mass.: G. & C. Merriam Company, 1981), s.v. "great plains cottonwood").

[79]Bennett, 13, says this was the second point at which the crossing was tried, the other being too crowded, and that there were fourteen wagons in all, i.e., a net gain of six since New Harmony, presumably Dexter's and Wilsey's parties.

[80]Bennett, 13, says that only part of the wagon's contents fell in the river; it also had Hamilton's, Lyon's, and Bennett's gear aboard.

[81]Independence, Missouri, now the first eastern suburb of Kansas City, well south of their starting point, St. Joseph.

Stewart, 131, says that this is 100 miles; William Fowler Pritchard's mileage reports add up to 138 miles, plus two days for which he reports no mileage.

[82]What species exactly is meant is unclear; see the entry for 2 June for a description of the animals he calls antelope. From this description, it appears that he is really referring to the only native American antelope – the pronghorn, *Antilocapra americana,* which has, in addition to the features described by William Fowler Pritchard, a white rump, grows to a maximum length of 4½' and a maxi-

mum weight of 140 pounds (Raymond E. Hall and Keith R. Kelson, *The Mammals of North America* (New York: Ronald Press, 1959), 339, hereinafter "Hall").

[83]The buffalo (properly, Bison) encountered here would have been *Bison bison athabascae,* or plains bison. *Bison bison bison* are the species found in the Rocky Mountains (strangely, these are known as "wood bison"). Bison are of the family Bovidae and may attain a length of 11½' and a weight of 2000 pounds. The native American wild ox, they have huge heads, a high shoulder hump, and short side fur and live almost entirely on grasses (although they will eat sagebrush when necessary). At this time there were tens of millions of them on the plains; they were reduced to a scant 541 in 1889 due to over-hunting and were only saved from extinction by captive breeding in zoos, most importantly the Bronx Zoo in New York City (Hall, 1023-26, Collins, 340, and William Bridges, *The Bronx Zoo Book of Wild Animals* (New York: The New York Zoological Society and Golden Press, 1968), 80). William Fowler Pritchard notes in the entry for 3 July that bison is the correct term for this animal.

Buffalo were seen here on 31 May for the first time (Bennett, 14). Later, on page 15, he contradicts himself and says he had seen neither buffalo nor antelope by 3 June. Stewart, 129, suggests that buffalo were not usually seen before the Platte River had been reached.

[84]Bennett, 14, says only one wheel needed repair.

[85]Bennett, 14, says that they saw dragoons on their way to Ft. Leavenworth, Kansas, on the Missouri River south of St. Joseph. Ft. Leavenworth was founded in 1827 by Col. Henry H. Leavenworth to protect travelers on the Santa Fe Trail *(The Encyclopædia Britannica, Micropædia,* 15th ed., s.v. "Leavenworth").

The wolves William Fowler Pritchard is likely to have seen or heard would have been *Canis lupus,* the grey wolf, not *Canis niger,* the red wolf. In the areas through which he went, the following sub-species would have been found: *C. lupus nobilus* in the plains, *C. lupus irremotus* in Wyoming and Idaho, and *C. lupus youngi* in Nevada and California. The grey wolf is bigger and has a heavier build than the red wolf (Hall, 847-52). The grey wolf may grow to 5½' in length and 100 pounds in weight (Collins, 318).

[86]Agriculturalists, living in log and earth lodges *(Webster's New International Dictionary of the English Language* (Springfield, Mass.: G. & C. Merriam Company, 1925) (hereinafter *Webster's),* s.v. "Pawnee").

The Encyclopædia Britannica, Micropædia, 15th ed., s.v. "Pawnee," says that they lived in dome-shaped earth-covered lodges on the Platte and had already ceded much of their land to the United States government in 1833 and 1848, and were to cede most of the rest in 1857.

Stewart, 129, says that the Platte was the border between the Pawnee to the north and the Cheyenne to the south, so they ought to be in Cheyenne territory. *The Encyclopædia Britannica, Micropædia,* 15th ed., s.v. "North American Plains Indians," says that the Pawnee lived south of the Platte.

[87]South of the aforementioned Big Blue River which it joins some twenty miles into Kansas. Bennett, 14, says this is the Republican Fork of the Blue River,

but the river in this vicinity which carries a name with this word in it is the Republican River which is a separate tributary of the Kansas River which is located to the west and south of the Little Blue River.

[88]W. F. Pritchard spells this "Carney." Bennett, 15, spells it "Kearney" and mentions that mail was sent to "the States" from it every two weeks. It is on the Platte River, ca. seventy miles ahead of their position at this point; the modern town is called Kearney (i.e., it has lost the "Fort" and gained a penultimate "e" which both Bennett and William Fowler Pritchard improperly included). It was named after Major General Stephen Watts Kearny (1794-1848), a major figure in the Mexican War (1846-48) who spent most of his life keeping order on the Great Plains (Encyclopædia Britannica, Micropædia, 15th ed., s.v. "Kearny, Stephen Watts"). Bennett, 14-15, doesn't mention the soldiers but says that game was very scarce due to the large number of emigrants although it was reputed to be abundant.

[89]The journal has "Republican Fork" lined out with "Little Blue" written in above; like Bennett (see note above for 2 June), he perhaps initially confused it with the Republican River.

[90]A major tributary of the Missouri River which runs all the way across southern Nebraska; W. F. Pritchard spells it "Platt."

[91]A member of the Vitaceae family of woody vines (Hylander, 369-73).

[92]Bennett, 15, says that one could see more than 100 wagons from the camp.

[93]Bennett, 15, says that they were 3 inches in diameter and 1½ inches thick.

Mushrooms are of the phylum Fungi, class Basidiomycetes. These plants must take in organic food — they cannot make organic material from inorganic material like most other plants. They are scavengers rather than parasites (i.e., they get their food from dead plants and animals, not living ones). They are usually found in damp forest habitats. The part of the mushroom usually seen is visible only during the reproductive phase — it is the platform from which the spores are launched. The main part of the plant are the mycelium, long filaments which are hidden underground (Hylander, 52-56).

[94]Lathyrus, similar to Vicia lathyroides (Merritt Lyndon Fernald, Gray's Manual of Botany, 8th ed. (New York: American Book Co., 1950), 930-32, hereinafter "Gray").

[95]Vicia sativa, or common vetch; an herb, not native to the United States, cultivated as fodder for livestock, hay, or winter pasturage (J. C. Th. Uphof, Dictionary of Economic Plants (Lehre: Verlag Von J. Cramer, 1968), 543). Related to the wild pea mentioned just above.

[96]Stewart says that this leg of the journey is 230 miles; William Fowler Pritchard's daily mileage reports add up to 258 miles.

[97]Nothing stands out here like the bluffs across from St. Joseph, and there is certainly no feature with this official name (i.e., capitalized as in William Fowler Pritchard's manuscript). The plains rise only about fifty feet above the Platte bottom (U.S. Department of the Interior, Geological Survey, 1:250,000 map for Grand Island, Nebraska).

[98]Bennett, 15, says this took place on 7 June.

[99]"A coarse cloth, usually printed with bright designs" but also in British usage "white cotton cloth;" William Fowler Pritchard's usage is unclear *(The American Heritage Dictionary,* s.v. "calico").

[100]Dragoons are medium cavalry – originally attacking on horseback and defending on the ground *(Encyclopædia Britannica, Micropædia,* 15th ed., s.v. "dragoon").

[101]Bennett, 15, says one mile back.

[102]Bennett, 15, says that these houses were frame houses, not sod houses.

[103]Bennett, 16, makes this identical observation on 12 June.

[104]Bennett, 15, specifies that this was mainly diarrhea.

[105]Bennett, 15, specifies Wilsey's party.

[106]This stream is too small to show on the *Nebraska 1985-1986 Official Highway Map.*

[107]W. F. Pritchard spells this "Nebrasca."

[108]It is a mile wide when full; it is named (from the French) for its extreme shallowness and is not navigable *(Encyclopædia Britannica, Micropædia,* 15th ed., s.v. "Platte River"). It is not the widest river in the United States since the Mississippi is as much as 1½ miles wide in places *(Encyclopædia Britannica, Macropædia,* 15th ed., s.v. "Mississippi River").

[109]Of the family Salicaceae which is composed of both shrub (poplars) and trees (willows), which are of genus *Salix,* perhaps from the Celtic, meaning "near water." This particular item is perhaps a Peachleaf Willow, a shrubby willow found in the central and western states (Hylander, 175). Gray lists 54 species of willows.

[110]Stewart, 129, says its normal average depth is only two feet.

[111]Bennett, 16, omits this important detail.

[112]Bennett, 16, specifies that this was Mills; at the time of the accident, he and Bennett were unloading wood.

[113]Bennett, 16, says 14 miles.

[114]The same remarks are made by Bennett, 16, for 11 June. This is a maximum of 95 miles from Ft. Kearny according to the (unnamed) guide book Bennett says they were using.

[115]Bennett, 16, says 10 graves, two of which were for ladies, aged 57 and 37 years.

[116]Of the family Pinaceae, which it shares with the Pines, the cedar is of genus *Cedrus* (Hylander, 148).

[117]Bennett, 16, mentions no names in connection with this ascent and apparently did not go on it himself.

[118]The following note appears (with several others) on the last two pages of book two: "Brady's Island. This Island took its name from the following circumstance. Three hunters and trappers were out on an expedition. Two quarreled when they went out to hunt. One was shot by accident as the other said, but the opinion was he was murdered by his companion. Hence its name."

[119]These forks are where the Platte divides into its North and South Forks.

[120]Called a "dog" due to the barking sound it makes. A social animal, living in underground "towns" formerly covering hundreds of square miles, but now virtually extinct in the wild due to extermination efforts of farmers and especially cattle grazers. It lives on green vegetation and short grass; thus its range was the western prairie where such grass was abundant. *Cynomys ludovicianus ludovicianus* or the black-tailed prairie dog was the most common and was the one seen here, in the north-south center of its range (Hall, 363-65). This animal is similar to the marmot and gopher which are mentioned just below, on 14 June, where the prairie dog and its town are described.

[121]Of the family Ranunculaceae, which grow in swamps and wet places (but not exclusively). These are herbaceous plants of a primitive type with generalized flowers. The genus *Ranunculus* ("little frog") has some 300 species and has petals (others of the family lack petals). Also known as Crowfoot (Hylander, 214-15).

[122]Of the family Cactaceae which has some 200 species in the United States, half of them in Arizona. Of the genus *Opuntia,* which it shares with the cholla and cane cacti; some 15-20 species of *Opuntia* are found in the plains states. It has flat pads and fruit which is eaten (Hylander, 311-20).

[123]Bennett, 16, mentions more graves (sixteen) seen, mostly of Missourians, a funeral, and an argument between Mills and Sweasey.

[124]Bennett, 17, adds that all had died of cholera.

[125]Bennett, 17, says that Dexter shot it with a revolver.

[126]Bennett, 17, says that they were carrying mail.

[127]Marmots are not the same as prairie dogs (see the note on prairie dogs for 11 June); marmots belong to genus *Marmota* or *Arctomys* and are commonly called woodchucks or ground hogs. They exist in both Europe and America. Both are burrowing rodents with bushy tails *(Webster's,* s.v. "marmot"). Since marmots do not live in villages, his statement that they are the same thing is inaccurate. Were he to have seen a marmot, it probably would have been a *Marmota flaviventris,* or yellow-bellied marmot, rather than a woodchuck *(Marmota monax)* which is found only in the eastern United States. The maximum length of this species is 19" and its maximum weight is 10 pounds (Collins, 278).

Bennett doesn't mention the prairie dog village at all.

[128]The owl is of order Strigiformes. Owls have large forward-facing eyes, acute hearing, a broad head with a facial disk, a short neck, and soft, fluffy plumage. They are silent-flying, largely nocturnal, birds of prey and feed mainly on rodents (Collins, 136).

[129]The gopher is another burrowing rodent like the prairie dog and the marmot but with a ratlike, smooth tail; of the family Geomyidae or of genus *Citellus,* a relative of the chipmunk; given the description of the mouth pouches, almost certainly the first *(Webster's,* s.v. "gopher"). A burrowing animal, with underslung jaws and cheek pouches; not social like the prairie dog; probably the northern or plains pocket gopher (Collins, 285-87).

[130]Bennett, 17, says that Sweasey found the ford, 3 miles below the usual one.

[131]This is the south fork of the Platte.

Stewart, 129, says it's 135 miles from the Platte to the South Fork of the Platte while William Fowler Pritchard's reported mileage is 175, plus 1 day without a mileage report. As for the preceding legs of the journey, his distances are longer than Stewart's.

[132]More so than the Mississippi or Missouri, according to Bennett, 18.

[133]The fork of the Platte is at an elevation of approximately 3280' while the Missouri-Platte fork is just below the 1000' level, so this description is reasonably accurate (taken from high points in the vicinity of these two locations from the *National Geographic Atlas of the World,* 6th ed.) The town of North Platte is at 2779' (U.S. Department of the Interior, Geological Survey, 1:250,000 map of Scottsbluff, Nebraska). The junction of the Platte is just below the 1000' contour (U.S. Department of the Interior, Geological Survey, 1:250,000 map of Omaha, Nebraska). The Platte is 310 miles (500 kilometers) long *(Encyclopædia Britannica, Micropædia,* 15th ed., s.v. "Platte River").

[134]Stewart, 117-118, indicates that it was usual for 1 man to be able to manage 30 head of loose cattle, usually with the aid of dogs. The entries for 16 June and 26 August indicate that the party had dogs but there is no indication that they were specifically used as herd dogs.

[135]For this day, Bennett, 17-18, gives an extensive description of the terrain of the Platte River, the bluffs, the water as a cause of illness, etc.

[136]Lavender is of the mint (Labiatae) family. If this is a native variety, it is probably of the genus *Hyptis,* the only American representative of some 300 tropical species. It is a fragrant whitish-green shrub with violet flowers. There are related species in swampy areas along the southeastern United States coast (Hylander, 452).

[137]These three plants are related. They are all of the group Compositae (Composites) and of the Thistle family (Carduaceae), a family of some 10,000 species of herbaceous and shrub types of plants, many of which are weeds. They are also all of the Chrysanthemum tribe which has conspicuous ray flowers and includes the daisies.

Wormwood is *Artemisia absinthium,* the herbaceous European plant which is the source of the French drink absinthe. *The American Heritage Dictionary,* s.v. "wormwood," says that it is also used as a tonic and in mothballs.

Camomile is *Anthemis* and the type found in the American West is Mayweed or Dog Camomile and has white or yellow flowers. Like wormwood, it is of European origin. Many of its species have medicinal uses. *The American Heritage Dictionary,* s.v. "chamomile, camomile," says that it is used for coughs and as a diaphoretic.

Sage is *Artemisia,* a shrubby plant with a bitter scent. There are some 200 species, 30 of which are common to the Central states. They have yellow or purple flowers and are a major cause of hay fever. The commonest type has silver-gray stems, wedge-shaped leaves, and yellowish-brown flowers. The type seen might have included *Artemisia ludoviciana* or Western mugwort or white sage (1-3 feet high) or *Artemisia tridentata* or sagebrush (3 feet high). This kind of "sage" is not to be confused with the herb sage, which is *Salvia occicinalis*, a mint, a half shrubby kind of plant, or some other kind of *Salvia* (Hylander, 470, 492-94, and Gray, 1523-24).

[138]W. F. Pritchard spells this "artmessia."

[139]Bennett, 18, says 23 miles done this day.

[140]Now a Nebraska State Historical Park between Ogallala and Oshkosh; a watering place in western Nebraska *(Nebraska 1985-1986 Official Highway Map)*.

[141]It is more likely that this was an alkaline flat; they are common throughout the area (personal experience of Mary Dohnalek).

[142]Bennett does not mention this remarkable feature of the terrain.

[143]Bennett, 18, says 23 miles.

[144]Not marked on the *Nebraska 1985-1986 Official Highway Map*. There is no such feature on the topographical maps of the U.S. Geological Survey. Moving upstream from Ash Hollow, there is a Signal Bluff at 6 miles and a Barn Butte at 12 miles. McCuligan Butte is opposite Oshkosh, and the last in this series is Coumbe Bluff, 6 miles due west of Oshkosh. Since William Fowler Pritchard mentions Courthouse Rock two days later and there is no bluff of any size between Coumbe Bluff and Courthouse Rock, "Castle Bluffs" must be one of these (U.S. Department of the Interior, Geological Survey, 1:250,000 map of Scottsbluff, Nebraska).

[145]Stewart, 130, confirms this.

The Sioux are now referred to as the Dakota. They formerly lived around Lake Superior but were forced out by other Indians before the white man came. At the time of William Fowler Pritchard's journey, they were fully adapted to a plains Indian style of life *(Encyclopædia Britannica, Micropædia,* 15th ed., s.v. "Dakota").

[146]Bennett, 19, says that there were two white men with them and about 20 people in all in the village.

[147]Bennett, 19, says that only one of the six tents was blown down – W. F. Pritchard was probably in the one that was, given his report of the incident.

[148]Bennett, 19, says 19 graves were seen; also that Dexter had to be talked out of leaving the train because the rest of the party stopped early on their own although he as scout was supposed to pick the stopping point. This is the first indication of friction between Dexter and the others and presages the split of the party on 21 July.

[149]Bennett, 19-20, quotes Palmer's description (Palmer was probably the author of one of the guides to the trail) and says that Beal and Bolton visited it. It is odd that he quotes someone else's description when he was in a position to give his own – he actually says that Beal's and Bolton's description was similar to Palmer's. He calls it Solitary Tower. Stewart, 130, calls this Courthouse Rock and mentions a Jail Rock which neither Bennett nor William Fowler Pritchard mention. The U.S. Geological Survey shows many named features in this area: Courthouse Rock is due south of Bridgeport, Roundhouse Rock is 5 miles due west of Courthouse Rock, Chimney Rock is south of Bayard, closer to the river than the two just mentioned; Castle Rock, Table Rock, Coyote Rock, Steamboat Rock, and Roundtop cluster 5 miles further along the river, and finally Scott's Bluff ends this collection

(U.S. Department of the Interior, Geological Survey, 1:250,000 map of Scottsbluff, Nebraska).

[150]Bennett, 20, says twelve o'clock.

[151]Bennett, 20, describes it as a haystack.

[152]This is not mentioned by Bennett.

[153]Of the family Culicidae; only the females bite (Donald J. Borror and Richard E. White, *A Field Guide to the Insects of America North of Mexico* (Boston: Houghton Mifflin Company, 1970), 266, hereinafter "Borror").

[154]Of genus *Pinus,* the pine is of the family Pinaceae, which it shares with the cedar (Hylander, 148).

[155]Bennett does not mention this entire area.

[156]Bennett, 20, lacks many of these details, mentioning only an Indian trader with a two horse wagon.

[157]The following note appears (with several others) on the last two pages of book two: "Scott's Bluffs. The name arises from a very melancholy circumstance which occurred some few years ago. There were some hunters out on an excursion. They had left the party and were lost. One was taken sick and could not travel any further. They were out of provisions and the sick man wished them to go and leave him which they did and some time after the bones of a human being were found near where he was left. The flesh had been torn off by the wolves."

[158]Bennett, 20, says there were 13 lodges and a row of huts.

[159]William Fowler Pritchard spells this "Pochahantas." She was Matoaka, daughter of Powhatan, who is said to have saved Captain John Smith's life by intervening with her father; Smith was one of the earliest colonizers of Virginia *(The American Heritage Dictionary,* s.v. "Pocahontas" and "Smith, John").

[160]Bennett, 20, says there were two Frenchmen, with Indian wives.

[161]Bennett does not mention this sighting.

[162]A tributary of the Platte River which runs south, mostly in Wyoming; the party is within a few miles of the present Nebraska-Wyoming border here.

[163]Summer solstice, normally around 22 June.

[164]This date conflicts with that found in a family Queen Anne Prayer Book *(The Order for Morning Prayer, Daily Throughout the Year)* of ca. 1708 which has a hand-written entry in a family genealogy which states that he was born on 13 June 1818, which is also the date passed down by family tradition. There is a baptismal record for 10 July 1818 which would leave a shorter time between his birth and baptism if the date in this journal is correct.

[165]Bennett, 20, says they went 18 miles.

[166]This tributary flows northward into the Platte River near Fort Laramie.

[167]Bennett, 21, says $1.50.

[168]A tribe related to the Dakota (Sioux), they were by this time located in Wyoming and heavily involved in the trading of guns and metal goods to the Shoshoni for horses. They were more-or-less continuously at war with the Blackfoot and Dakota and as a result sided with the white man during the later Indian wars *(The American Heritage Dictionary,* s.v. "Crow").

169Stewart, 130, says this leg is 130 miles; William Fowler Pritchard says 142 plus one day without mileage reported (plus Bennett's figure has been used for one other day for which William Fowler Pritchard makes no report).

170This data (with the exception of the number of deaths) is repeated (with several other unrelated notes noted elsewhere) on the last two pages of book two.

171Bennett, 20-21, has vastly less detail for this day.

172Tar is what is produced by the destructive distillation of various organic substances, primarily woods and coal. It has some medicinal uses but it is unknown what exactly is meant by "extract of tar" *(Encyclopædia Britannica, Micropædia,* 15th ed., s.v. "coal tar" and "wood tar").

173Bolton and Axton fetched the wagons, according to Bennett, 21, who does not mention W. F. Pritchard's participation in this excursion, perhaps because he did not own one of the wagons. Nothing further is known about Axton.

174Bennett, 21, says that this was the Thompson party, and that it had 200 loose cattle and 500 cattle in all.

175Bennett, 21, calls it "Warm Spring."

176A tributary of the North Fork of the Platte which runs west; *Wyoming 1987* (state highway map) omits the "Bitter" part of the name.

177Another version of this day's events is found (with several other unrelated notes noted elsewhere) on the last two pages of book two: "27 June – Bitter Cottonwood Creek. This part of the creek runs through a narrow valley, each side bound up with walls of rock. This was a few years ago a great place for the Indians to attack emigrants but the number keeps them in check and the fear of the smallpox, a disease they are afraid of, it having swept off whole tribes of them. A good thing it keeps up?"

178W. F. Pritchard spells this "Larime;" it is 10,274' high *(Wyoming 1987* (state highway map)). Bennett omits this sighting.

179See note on Horseshoe Creek just below.

180Willow Creek joins the Platte from the east and does so *after* Horseshoe Creek, which joins the Platte from the west. Bennett, 21, spells this "Horse Shoe Creek." It is about 14 miles from Cottonwood Creek. William Fowler Pritchard seems to have his geography a bit mixed up. There are a number of creeks with the same names and it is possible that there was a different Willow Creek back in 1850. Today, both of these creeks flow into Lake Glendo, created by a dam 10 miles upstream from Fort Laramie *(Wyoming 1987* (state highway map)).

181The elk is the Wapiti or *Cervus canadensis,* a member of the same family (Cervidae) as the deer. Also called the American Elk, it is the largest of the family in North America, growing to a maximum size of 9½' and a maximum weight of 750 pounds, and prefers semi-open woodlands (Collins, 336).

182Either mule deer, *Odocoileus hemionus,* or white-tailed deer, *Odocoileus virginianus,* which, as its name suggests, is the more common in the eastern part of the United States while the mule deer is more common in the west. At this point, it could be either but is more likely to have been a mule deer (Collins, 336-39).

93

He makes an almost identical remark about the variety of local fauna on 2 July, but only explicitly says that he saw any deer on 21 September, while in the Sierra Nevadas, near the end of the trip.

[183]This is most likely Elkhorn Creek, which is about 20 miles upstream from Cottonwood Creek *(Wyoming 1987* (state highway map)).

[184]Bennett, 21, says they went through "the Black Hills" and up Bitter Cottonwood Creek.

[185]Bennett, 22, names only LeBonte or Big Timber Creek which is the most notable creek they would have passed over in this stretch, so this is likely another name for this creek which William Fowler Pritchard had heard.

[186]W. F. Pritchard and Bennett call it La Prele *River.* Bennett, 22, seems to say that the two groups rejoined one another before this point. This is another northward flowing tributary of the North Fork of the Platte *(Wyoming 1987* (state highway map)).

[187]This is a lizard, not a toad, *Phrynosoma douglassi,* which reaches up to 4" in length and is virtually hornless with a short tail – the type which lives further south has more prominent horns (Collins, 370).

[188]There are several families of grasshoppers; they are all of the order Orthoptera to which crickets and cockroaches also belong. Most species have wings and are large; only the males "sing" by rubbing their legs together. Just which grasshopper is meant here cannot be determined (Borror, 76).

[189]Possibly John Beal (1797-1863). One of the "Boatload of Knowledge" emigrants brought to New Harmony in 1826 by Robert Owen. Beal participated in David Dale Owen's surveys and taught cabinetmaking at the School of Industry. He married Roxane Clarke; their daughter Caroline was born in 1825 near Pittsburgh while the family was en route to New Harmony on the Boatload of Knowledge; she died on 1 May 1894 in New Harmony. John Beal had at least two daughters, one of whom (Caroline) married (in 1847) Adam Lichtenberger, who is mentioned in William Fowler Pritchard's letter to his wife, below. The Lichtenberger card mentions that John Beal's wife was Roxane Clarke (Local History Card file).

John Beal was also probably the father of George Beal, who William Fowler Pritchard's grandson Thomas MacDonagh Pritchard knew (information from him via Thomas Pritchard's son, Earl H. Pritchard).

[190]W. F. Pritchard's text seems to be "Otyman." The Local History Card file has a card for a "William Oatzman." He is described as having bought the Bolton Carding Machine in 1847 in New Harmony (see note for 20 May on Samuel Bolton). The Foreword to Bennett uses the form "Oatzman" also, while Bennett's text itself has him as "Otzman" (28 April and 1 and 22 July) and also "Fotzman" (2 July, presumably a typo). If William Fowler Pritchard used an alternate form of "z" which has a tail like "y," "g" and "j," the name might be interpreted as "Otzman" without an "a" and this is the form which has been used.

[191]They were found the next day – see below. Bennett, 22-23, reports this event on 1 July but confirms that it happened 30 June.

[192]Both gooseberries and currants are of the genus *Ribes* and are bushy plants (Hylander, 266-68). Wild gooseberries are *Ribes cynosbati* and are prickly (Uphof, 452). Not all currants are red; William Fowler Pritchard himself mentions both red (14 July and 1 August) and yellow (1 August) varieties.

[193]W. F. Pritchard later spells this "Fourche Boise," "fourche" meaning divided or forked. Bennett, 23, calls it Fouche Boise *Creek*.

It is today called Boxelder Creek *(Wyoming 1987* (state highway map)), or Box Elder Creek (U.S. Department of the Interior, Geological Survey, 1:24,000 quadrangles for Glenrock and Careyhurst, Wyoming). This creek very definitely does not originate in a spring five miles up in the hills – it originates a considerable distance south of where they encountered it.

[194]There is a blank space in the journal before "feet." Perhaps William Fowler Pritchard intended to get this information and never did. They are at approximately 5200' (U.S. Department of the Interior, Geological Survey, 1:24,000 quadrangles for Glenrock and Careyhurst, Wyoming).

[195]Of genus *Mentha*, of which there are 12 species (Hylander, 452). The mints are the family Labiatae, to which lavender also belongs (see note for 17 June)

[196]Thyme is of the mint family (see note just above), of genus *Thymus*, specifically *Thymus serpyllum*. It is of European origin, a low matted growth with purplish flowers (Hylander, 453, Gray, 1245).

For this date, Bennett, 23, also mentions an argument between himself (supported by Otzman) and Sweasey over some sugar and hard bread they had purchased which Sweasey refused to put into the wagon on the "frivolous pretense" of too much weight. This suggests that these two, and probably William Fowler Pritchard as well, were tied to Sweasey's party primarily because they had arranged for wagon space for their belongings in Sweasey's wagon.

[197]Actually, somewhat higher – see note on Boxelder Creek above.

[198]The beaver is *Castor canadensis,* which grows to a maximum length of 25-30" (exclusive of a tail of up to 10") and attains a maximum weight of 60 pounds (Collins, 288-89).

Bennett does not mention these details about the beaver the party saw.

[199]William Fowler Pritchard is probably referring here to the black bear, *Euarctos americanus,* rather than the grizzly, *Ursus horribilis,* because he specifies "grizzly" in the entry for 19 September. The black bear attains a maximum size of 5½' and a maximum weight of 350 pounds while a grizzly may reach 7½' and 500 pounds (Collins, 320-21).

[200]Bennett does not mention these details.

[201]There are two Muddy Creeks in the vicinity today, neither of them called "Crooked." The larger, just "Muddy Creek," comes first, about 10 miles beyond Deer Creek, and flows into the Platte from the south. The second and smaller, "Dry Muddy Creek," joins the Platte from the south some 6 miles further on. Likely the larger is intended *(Wyoming 1987* (state highway map)).

[202]One of the few places W. F. Pritchard uses "the" with "Platte."

[203]Bennett, 23, says nine miles.

[204]A twilled cotton cloth – from 'Genoa' *(Webster's,* s.v. "jean, also jeans").

[205]Bennett, 23, says that among those attacked (and getting a share of the meat when it was killed) was Jackson.

[206]Hares (also known as jackrabbits) differ from cottontail rabbits in that they are larger and have longer ears and hind legs; cottontails are *Sylvilagus,* hares *Lepus.* The ones seen here were probably white-tailed jackrabbits *(Lepus townsendii),* which are 18-22" long (Collins, 274).

[207]I.e., recuperate *(The American Heritage Dictionary,* s.v. "recruit").

[208]Bennett, 24, says that Pullyblank was the butcher.

[209]Bennett, 24, went on this expedition and describes a five hour ascent of a mountain.

[210]*Ovis canadensis canadensis,* which has thick horns which are curved nearly 360° and which may weigh up to 343 pounds (Hall, 2:1031).

[211]*Acer negundo,* of the maple genus, this is a maple with compound leaves, also known as the Ash-leafed maple (Gray, 988). It grows to 70' in rich, moist soils and withstands drought and cold well but has poor wood (Hylander, 355).

[212]An unidentifiable variety; see note on cottonwoods for 28 May.

[213]Bennett, 24, says that this was a Mormon ferry and was charging $5 per wagon and that the leaders decided to avoid it, against the wishes of the company.

[214]They are now very near present-day Casper, Wyoming.

[215]Stewart, 130, says this leg of the journey is 130 miles; William Fowler Pritchard reports 107½ miles, plus one day with no mileage report. They didn't use the "regular ferry," and perhaps this accounts for this.

[216]Bennett, 24, says that these two were O'Neal (he doesn't specify whether it was John or Mitch) and [William] Faulkner and that one wagon had to go across the river with its springs and axles still attached due to its construction and that it was thus too heavy. Stewart, 131, mentions that the loss of animals and even people was not unusual.

[217]Bennett, 24, says that W. J. Sweasey was one of these men. William Fowler Pritchard says on 8 July that the wagon which went downstream belonged to Mr. Dexter.

[218]Bennett, 24, says 11 o'clock.

[219]Bennett lacks this description of Mrs. Dexter's travails.

[220]This remark is hard to reconcile with his statement for the previous day that they avoided the regular ferry and used a "fixing" another group had rigged. Perhaps they used the ferry for the last few wagons for some reason not mentioned – perhaps because of almost losing some of their party at the "fixing" earlier in the day.

[221]Bennett, 24, says 22 cattle.

[222]In Shropshire, England, where he was born. See the Introduction.

[223]Combs brought 10 of them in near sunset and Williams the rest (Bennett, 24).

[224]Bennett, 23, mentions this dissatisfaction earlier (on 6 July).

[225]Bennett, 24, does not mention the exception.

[226]Nothing more is known about this person. See 18 August 1850 for a report about the finding of his body.

[227]Samuel and William Bolton; see remarks on 21 July.

[228]On the south side of the Platte, about 10 miles downstream from Casper.

[229]Bennett, 25, says 10 miles.

Their camp site is Poison Spring Creek which flows eastward into the North Fork of the Platte just south of Casper (U.S. Department of the Interior, Geological Survey, 1:24,000 quadrangle for Clarkson Hill, Wyoming).

[230]This "Rock Avenue" is just above Poison Spring Creek. The word "descent" is deceptive: they would have been going up it, not down it. The map does not give it a name (U.S. Department of the Interior, Geological Survey, 1:24,000 quadrangle for Clarkson Hill, Wyoming).

[231]The spring is the source of Willow Creek (altitude 6199'), which they have been following closely since coming up 'Rock Avenue' (U.S. Department of the Interior, Geological Survey, 1:24,000 quadrangle for Benton Basin NE, Wyoming)

[232]This is now called Ryan Hill, altitude 6486', and gives the best view backwards over the area they have just traversed – at its top, the road continues on over a plateau which slopes very gradually down into the Sweetwater River valley (U.S. Department of the Interior, Geological Survey, 1:24,000 quadrangle for Benton Basin NE, Wyoming).

[233]No range with this name appears today on the *Wyoming 1987* state highway map; he is probably referring to the Rattle Snake Mountains, some 15-20 miles due west of where they are at this point.

[234]Bennett, 25, says 32.

[235]Bennett, 25, reported that he was unable to walk.

[236]*Sarcobatus vermiculatus*, or creosote *(The American Heritage Dictionary,* s.v. "greasewood"). Sarcobatus is one of the Goosefoot or Pigweed family (Chenopodiaceae) and is a thickly branched spiny desert shrub (Hylander, 204-5). There is also a member of the rose family with this name (Hylander, 296, 304) and absinthe is a member of the Thistle family (see entry for 17 June for several other members of this family which he saw).

[237]Bennett, 25, says that there was one dead and one missing when Sweasey went back the next day to look.

[238]*The Oxford English Dictionary,* 2nd ed. (Oxford: at the Clarendon Press, 1989) (hereinafter *OED2),* s.v. "lalescitors," says 2NaO.CO (normal sodium carbonate). William Fowler Pritchard capitalized this word.

[239]Bennett, 25, says "Sweet Water."

[240]Bennett, 25, says 5 miles.

[241]He repeats his description made just above, but with doubled sizes!

[242]Bennett, 25, says 10 miles.

[243]Including the sick Bennett (Bennett, 25).

[244]Stewart, 131, says this leg is 50 miles, William Fowler Pritchard 42 (but one day had no mileage reported).

Thousands of names of emigrants are inscribed on this rock.

[245]Bennett, 25, says 70 or 80' high.

[246]Bennett, 26, says that he made the ascent on the northwest side with several others, but makes no mention of William Fowler Pritchard's doing it. Stewart, 131, says almost everyone passing this point did the climb.

[247]About 5 miles due southwest of Independence Rock (U.S. Department of the Interior, Geological Survey, 1:100,000 30x60' map of Bairoil, Wyoming).

[248]Stewart, 132, says the wagons go through via an easy pass in the hills about half a mile to the south.

Bennett does not mention this but notes that the road goes through another defile "a considerable distance to the south."

[249]Note that this is contradicted at the end of this day's entry where 16 miles is specified. Bennett, 26, says 12 or 13 miles.

[250]Bennett, 25, says two or three hundred yards.

[251]Bennett, 25, does not mention that one man was a cripple and says that the man William Fowler Pritchard said was a cripple came up to them in a state of great agitation, asking for some gunpowder, which they refused.

[252]Bennett, 26, says eight of these were Dexter's, two Sweasey's.

[253]Today these are called the Granite Mountains. They are abrupt and high for the area and the river comes up to their southern edge; Devil's Gate is their southeastern terminus (U.S. Department of the Interior, Geological Survey, 1:100,000 30x60' map of Bairoil, Wyoming).

[254]A steplike configuration in igneous rocks *(Webster's,* s.v. "trap, also traprock").

[255]These types of snake are mostly tropical and are found mostly in the southeastern United States, but this variety, *Farancia abacura reinwardtii,* is found in the Mississippi valley, presumably including this very distant tributary. They can be up to 7' long and 2" in diameter, are purple black on top, vermilion with black blotches on the sides and bottom, and are called hoop snakes because of the myth that they can take their tails in their mouths and move by rolling along like a wheel – they can't but can be found in a circular configuration resting on the ground (Ditmars, 122-24).

[256]This is probably what is known today as Split Rock – the spur of the Granite Mountains which comes closest to the river (U.S. Department of the Interior, Geological Survey, 1:100,000 30x60' map of Bairoil, Wyoming).

[257]Bennett, 26, says 13 miles.

[258]Bennett, 26, says 13 miles.

[259]Sodium or potassium bicarbonate (i.e., baking soda) *(Webster's,* s.v. "saleratus").

[260]William Fowler Pritchard says "snowcaps" (also below).

[261]Bennett, 26, says these were sighted on 16 July.

[262]The major range of western Wyoming, running from northwest to southeast and containing the South Pass over which the trail passed at their southern end.

263Bennett, 27, says 10 o'clock.

264Stewart, 132, calls this "Ice Slough;" he also mentions Split Rock and Sweetwater Rocks which William Fowler Pritchard does not. Bennett does not mention Ice Springs. There is an "Ice Slough" on the 1:24,000 map for the Myers Ranch, Wyoming quadrangle (U.S. Department of the Interior, Geological Survey). It is, however, a good ways back on the trail from where they camped this day, on the part of the trail which leaves the Sweetwater and runs to its south, so if this is the place meant, it is mentioned out of order.

265Bennett, 27, says they travelled 7 miles and that they found sage which was six feet high and six inches in diameter.

266Bennett, 27, does not mention this but says that this was the roughest stretch of road they had travelled.

267Bennett does not mention this.

After mentioning being snowed upon, it seems contradictory to say that it was the warmest day since they left Fort Laramie, and that the night was cold. Perhaps it was extraordinarily cold for this period of the mid-summer in this area this year. If daytime snow is "warm," what must a "cold" night be like?

268Bennett, 27, says that he was one of the four guards and that they went 2 miles.

269Bennett, 27, says 7 or 8 miles.

270Bennett mentions none of this.

271The trail just before the original South Pass (see following note) passes between two hillocks which are about 20-30 feet higher than the trail; one is circular, the other oblong and they are not named on the map (U.S. Department of the Interior, Geological Survey, 1:24,000 quadrangle for Dickie Springs, Wyoming).

272The *Wyoming 1987* state highway map says South Pass is at an elevation of 7,550', but this is the new South Pass on Wyoming state highway 28, not the South Pass over which the emigrant trail passed, which is to the south of this and at an elevation of 7412' (U.S. Department of the Interior, Geological Survey, 1:24,000 quadrangle for Pacific Springs, Wyoming). Stewart, 132, says that this part of the trip is 100 miles, which is almost the exact figure reported by William Fowler Pritchard – 98½ miles.

273This spring is the source of Pacific Creek which joins Big Sandy which in turn joins the Green River (U.S. Department of the Interior, Geological Survey, 1:24,000 quadrangle for Dickie Springs, Wyoming). Since the Green River joins the Colorado, "Pacific" is technically correct, but since the Colorado empties into the Gulf of California which is separated from the Pacific Ocean by Baja California, this is a bit misleading – it's not as if the water they see here is going in the same direction they are.

274The Green River is a major river which is the southward drainage of the Wind River Mountains. It runs north-south across the southern part of western Wyoming and the eastern part of north and central Utah, joining the Colorado River in southeastern Utah.

275William Fowler Pritchard was a cabinetmaker who later was New Harmony's undertaker and built a hearse – see the Introduction.

276This major event of the trip is poorly explained here; Bennett, 27, says that the final issue which brought the train to split up was whether to travel all night or to do the stage in two days without feed or water. Williams, Dexter, and Pullyblank wanted to cross the next stretch in one night while Sweasey wanted to do it in two days.

277John Williams was born in 1818 and was married to Lydia Hoover (1803-88) and is listed as a butcher in the business ledger of Sampson & Fauntleroy (Local History Card file). The California Argonauts 1850 card lists him and suggests that his wife was the daughter of John Corbin [Bennett mentions a Corbin on the trip]. Other relatives may be mentioned on this card; it is unclear from the way the entry is worded whether they are Williams or Corbins. His marital status is mentioned in William Fowler Pritchard's letter to his wife, below. Bennett does not mention him.

William Fowler Pritchard's manuscript has some odd spellings here: "Decstor" for "Dexter," and "Pulabank" for "Pullyblank."

278Nothing more is known of Mitch O'Neal from the Local History Card file, except that he perhaps had a brother named John – this is mentioned in William Fowler Pritchard's letter to his wife, below. Bennett does not mention him.

279Bennett, 28, says that they joined Bolton and Hinkley and went on to join Dexter's party on 22 July.

280Bennett's description makes out that the all-night party (Dexter, Wilsey, and Pullyblank (he does not mention Williams)) left on the 21st, presumably leaving everyone else with the original leader, Sweasey, the leader of the two-days party. They were joined according to him by Messrs. Beal, Otzman, Bolton, and Hinkley (who William Fowler Pritchard does not mention) on the 22nd. Sweasey's party now consisted of at least himself and his family, William Fowler Pritchard, Bennett, Mills, Moore, Fewer, Jackson, Lyon, and Spencer.

281Bennett, 28, does not mention this.

282A southward flowing tributary of Big Sandy Creek, itself a tributary of the Green River. The Little Sandy flows into the Big Sandy about 20 miles from the Green River.

283Stewart, 133, and William Fowler Pritchard agree on the mileage for this leg of the journey (i.e., 20 miles).

284Bennett, 28, says 3 o'clock.

285Only a few miles west of the Little Sandy (on which, see the note for the previous day).

286Bennett, p 28, says 4 miles.

287John Charles Frémont (1813-1890), surveyor and explorer of the West who played a questionable role in California during the Mexican War; first senator from the state of California, first Republican candidate for President (1856), and governor of Arizona Territory 1878-1883 *(Encyclopædia Britannica, Micropædia,* 15th ed., s.v. "Frémont, John Charles").

[288]Actually, 13,745' high *(Wyoming 1987* (state highway map)). They are some 54 miles due south of this peak at this point.

[289]Bennett, 28, says that the coming stage is "Soublett's cutoff" which is 40 miles long without water and which was done continuously. Stewart, 133, agrees; the route prior to 1841 was south to Ft. Bridger, then north again. Sublette's cutoff was normal from 1844 onwards.

[290]Bennett, 28, says 5 o'clock.

[291]Sage fowl are not mentioned by this name in *Webster's,* however sage grouse are. They are large grouse of species *Centrocercus urophasianus* (s.v. "sage grouse").

Bennett, 28, was one of these hunters and he says there were 4 large sage fowls and one hare.

[292]This is not mentioned in Bennett.

[293]The "Old Mormon Ferry" just below La Barge on Fontenelle Reservoir which was then the Green River *(Wyoming 1987* (state highway map)).

[294]Bennett, 28, says 30 at breakfast. Stewart, 133, says 45 miles but this area has the widest variety of estimates.

[295]By Bennett and four others (Bennett, 28).

[296]Bennett, 28, says the cattle were driven to camp starting at sunrise, until 10 o'clock.

[297]Bennett, 28, says there were 20. Stewart, 134, says this is a Snake-Sioux borderland – they fought each other, not the emigrants.

The Encyclopædia Britannica, Micropædia, 15th ed., s.v. "Shoshoni," says that most Shoshoni speakers were typical plains Indians, but that the Diggers, mentioned below on 22 August, were also Shoshoni and lived a very different lifestyle.

[298]Queer, odd; derogatory *(Webster's,* s.v. "rum").

[299]William Fowler Pritchard's description of these Indians here is much more detailed than that of Bennett, but Bennett describes them in greater detail the next day at their big encampment.

[300]Bennett, 28, ways 10 to 13 miles. Their camp was probably on Fontenelle Creek, south of where they crossed the Green River (which is now under Fontenelle Reservoir) *(Wyoming 1987* (state highway map)).

[301]Trout are of the family Salmonidae, are carnivorous, and live in cool water. The species seen by William Fowler Pritchard was probably the Cutthroat Trout, *Salmo clarki,* which is commonly 15" long and the record weight for which is 41 pounds (Collins, 524).

[302]Bennett, 29, says at least 500.

[303]Family Ostreidae, genus *Ostrea;* a marine bivalve mollusk (see also note on mussels in the entry for 16 August) *(The American Heritage Dictionary,* s.v. "oyster" and Collins, 591).

They are a few miles northeast of what is now Fossil Butte National Monument; the whole area is rich in fossils.

304Bennett, 29, says that their maximum size was 9 inches.

305William Fowler Pritchard says "steeped." Stewart, 134, agrees, terming these roads the worst yet.

306Bennett, 29, says seven and a half miles.

307This runs through Kemmerer a few miles south of where they are and then on for some 50 miles before joining the Green River near Rock Springs.

308Bennett, 29, says Canadian.

309*Webster's,* s.v. "backhand," says this word is used for the band holding the sheets of a book together as well as in the specific sense in which it is used here.

310Here again William Fowler Pritchard has more details than Bennett.

311Bear River Divide, next to the Utah border in southwest Wyoming, near Cokeville *(Wyoming 1987* (state highway map)).

312Bennett, 30, says 15 miles. He also mentions the view from the summit and roads that were even worse than those passed during the previous days.

313Another southward flowing tributary they had to cross, then up and over the ridge between it and the Bear River; while doing this they passed from Wyoming into Idaho.

314North of Cokeville, Wyoming, just across the border from Idaho. The Bear River begins in the easternmost part of the Utah panhandle, runs nearly north well into Idaho, then changes direction to nearly due south and drains into the Great Salt Lake. The wagon route runs north, up the river to its northernmost point, then leaves it heading west.

Bennett, 30, does not mention Smith's Fork and says that they encamped on the Bear River only after travelling 10 miles the following day.

315Stewart, 133, says this leg is 110 miles long; William Fowler Pritchard says 162, but there are two days of unclear "mileage" on the trail in his account.

316Now in Idaho, this is the second major southward flowing fork they had to cross.

317See *Webster's,* s.v. "a," on its use as a preposition.

318Belonging to the family Catostomidae, who are bottom feeders (Collins, 531). The Buffalofish, mentioned on 25 May, are of the same family.

319*Ambloplites rupestris,* i.e., redeye, of the family Centrarchidae, which also includes sunfish. They grow up to 1' in length and weigh up to 1 pound; they are found in the Mississippi (Collins, 550, 555).

320Bennett, 30, does not mention this name.

This is probably a reference to the Preuss Range as a whole – the canyon through which they go after reaching "the top" is probably Montpelier Canyon. There's a not very large bump on the way to this canyon which they would have had no reason to go over rather than around, and the Geneva Summit is a bit further on. This route is a difficult one they have taken to save a few miles detour over easier ground to the south (U.S. Department of the Interior, Geological Survey, 1:250,000 map for Preston, Idaho).

321Bennett, 30, says that this is the worst place until the Sierra Nevadas.

322There are a great variety of crickets, including several which attain this size; it is impossible to know which variety William Fowler Pritchard saw. Crickets are of the family Gryllidae and are related to the grasshopper (a variety of which he mentions in the entry for 30 June) (Borror, 82).

323The text looks like William Fowler Pritchard says "spring lance." Bennett, 31, clearly names a place "Spring Branch."

324Bennett, 30, mentions a Mr. Moore and some Canadians having a near fatal encounter with the Bear River, taking dinner at a place called Big Timber, and camping at Spring Branch.

325Bennett, 31, says 20 miles.

326Bennett, 31, does not mention this but mentions three kinds of wild currants.

327Bennett, 31, mentions going through Soda Springs (which is misprinted "Loda") which William Fowler Pritchard mentions for the next day.

320Now a town of that name, elevation 5779', population 4051 *(Idaho Official Highway Map,* 1985). This is the northernmost point of the Bear River. Bennett, 31, refers the reader to Frémont's description. Stewart, 134, also mentions a Beer Spring.

329I.e., 165 by 66 feet; one rod equals sixteen and a half feet *(The American Heritage Dictionary,* s.v. "rod").

330Cypress is of the family Cupressaceae and of the genus Cupressus, which is represented by 12 species in the United States. It is a conifer like the pines, has small spherical cones, is aromatic, and has twigs which are closely covered with minute overlapping scales (Hylander, 168-70).

331There is a "Steamboat Hill" just to the west of the present day town of Soda Springs, on the north side of the river (which is now Alexander Reservoir). It rises some 200' above the surrounding landscape, so this cannot be the "Steamboat Springs to which William Fowler Pritchard refers. The individual springs of this region are by and large not named on the U.S. Department of the Interior, Geological Survey, 1:24,000 quadrangle for Alexander, Idaho, so we can only suspect that this is where the spring of the same name was located before the reservoir was built.

332This is the city through which William Fowler Pritchard first entered the Unites States in 1842; he mentions that he lived there for a time at the conclusion of the journal. See the notes to the Introduction for further details.

333This is the road to Fort Hall, the normal route followed until this year.

334Probably Steamboat Hill, mentioned several notes earlier.

335Bennett, 31-32, says that two of the party visited this hill and found it to be 300 *yards* across and 60 feet deep.

336Bennett, 31, says that they stopped at 9 p.m.

337Bennett, 31, mentions that this is a road junction: on the right going to Fort Hall, on the left (which the party took) Myer's or Headgpeth's (or Hudspeth's) cutoff.

J. S. Holliday says that this cutoff (also called "Emigrant's Cutoff") was *regular* in 1849 and the only part of the California Trail to be pioneered by gold rush travelers (J. S. Holliday, *The World Rushed In: The California Gold Rush Experience* (London: Victor Gollancz Ltd, 1983), 203).

[338] Bennett, 31-32, states that a party of 5 went ahead looking for water, found this note but then encountered Mr. Moore (who was with the other party) who pointed out a nearby stream which is the one they went to for breakfast. This is one of the real discrepancies between William Fowler Pritchard's and Bennett's accounts. This stream, the one mentioned as their campsite for the day, or one of the streams mentioned the following day may be one of the two branches of the Portneuf River which runs north into the Snake River.

[339] Bennett, 32, says 10 miles.

[340] Bennett, 32, mentions that they encountered a stupendous view for a painter.

[341] Ducks and geese are of the family Anatidae, which they share with swans. They are medium to large birds who live on both fresh and salt water, have webbed feet (note William Fowler Pritchard's comparison of their feet to those of beavers in the entry for 2 July), bills with tiny sawteeth along the edges, and build nests of vegetation on the ground near water. There are many species and it is impossible to say which are meant here (Collins, 31).

[342] W. F. Pritchard spelled this "Colombia," like the South American country.

[343] Bennett, 33, mentions that he was sick and thus could not hunt the many water-fowl they saw near this swampy stream but that Mr. Moore shot a dozen ducks, mainly blue winged teal, and Mr. Fewer shot 4 or 5 near camp.

[344] This paragraph and the one following it are both dated August 7th. By comparing the events with those reported by Bennett, 33, the dates of the following entries have been adjusted (up to but not including 10 August).

[345] Bennett does not mention this.

[346] The spelling of this word is very hard to make out. The second letter is certainly "o", the fourth and fifth "et", the others appear as follows: the first is "T" or "F", the third "s" or "r", the sixth "e" or "o", and the seventh "p" or "th". After research, the version used was selected due to the existence of a Toxteth Ward and a Toxteth Park Cemetery Workhouse in *Mawdsley's Map of the City of Liverpool and Suburbs, 1884*. For William Fowler Pritchard's connection with Liverpool, see the notes to the Introduction.

[347] Bennett, 33, mentions that they met a trading party going from Salt Lake to Fort Hall and that the turnoff to Salt Lake was 15 miles ahead at that point.

[348] Reported by mistake as 7 August (as was the preceding entry – see note for it).

[349] Stewart, 137, says that these Indians were then known only as Diggers – but William Fowler Pritchard's use of "Shoshones" disproves this assertion.

[350] Bennett, 34, notes that they we at the Salt Lake cutoff.

[351] Reported by mistake as 8 August (see note for 7 August entry).

[352] Bennett, 34, says 18 miles.

353This appears to be a second entry for the same day.

354Bennett, 34, says 17 miles.

355Bennett, 34, says 4 miles.

356Bennett, 34, says 8 miles.

357Bennett, 35, gives a long description of their problems with the dust for this day.

358A tributary of the Snake river to its south. Most of Hudspeth's Cutoff goes along the line which divides the Great Salt Lake from the Snake River drainages – here they have to cross a river which drains northward into the Snake.

359Dull brownish yellow or yellowish gray; also a kind of cloth. The exact type of bird referred to cannot be determined from the information given.

360Bennett, 35, says 20 miles. Stewart, 135, shows a gain of only 5 miles! William Fowler Pritchard makes the length of the cutoff itself 142 miles, Stewart 130. Note the remark for the next day about the number travelling with them, indicating that the Fort Hall route was the most popular, as Stewart says it was for all years *except* 1850, when the route taken was the most travelled – see the Introduction.

361This may be a place named "The Narrows" on the 1:24,000 quadrangle for Chokecherry Canyon, Idaho (U.S. Department of the Interior, Geological Survey). The Raft River at this point goes through the southern tip of the Jim Sage Mountains with a 700' high hump to the south of the river; the Raft River has unexceptional scenery both before and after this point, so it is likely that this one mention of a mountain is this place. The map labels the trail here the alternate route to Salt Lake City, but it is also Hudspeth's cutoff.

362Bennett, 35, says 13 miles.

363This is short for Aspen, which are of the family Salicaceae (willows), and of the genus Populus (poplars). Aspens have leaves with flattened leaf stalks, which make them "quake" in slight breezes (Hylander, 176). It is of the species *Populus tremuloides,* with various sub-species in the West (Gray, 521-22).

364Of the Poaceae (grass) family, which has over 7,000 species. Of the oats tribe, genus *Avena.* Oats were of Eastern European origin and were not cultivated widely in Western Europe until the Middle Ages. Oats are mostly used for stock fodder and thrive in a cool, moist climate. A specifically silver variety is not mentioned (Hylander, 503, 507-8).

365Of the family Linaceae (of the Geranium order), genus *Linum.* The family has over 150 species (of both herbaceous and shrubby types) with simple, narrow leaves and five petalled flowers. *Linum* is the long-cultivated ingredient of linen and the source of linseed oil (Hylander, 338-39).

366Just what grass this is cannot be determined.

367Bennett, 35, says it's called Pyramid Valley – William Fowler Pritchard did not "name" it. Bennett remarks that one boulder where they stopped for lunch was 300' high and inscribed with names. Holliday calls this "Steeple Rocks" while the Idaho Official Highway Map calls it "City of Rocks," a California Trail

Landmark. They are just north of the Utah border here, and less than twenty miles east of the Nevada border.

The U.S. Department of the Interior, Geological Survey 1:24,000 quadrangle for Almo, Idaho, contains a "City of Rocks" which is about 3½ miles west southwest of Almo. It also shows the trail going just to the east of the area, which is a "glen" much as William Fowler Pritchard describes it. The named rocks include Register, Camp, Treasure, Turtle, Bath, and Camel – the one named Register Rock is likely the one Bennett describes as inscribed with names. There is a place named Castle Rocks some three miles northwest of Almo, but the trail probably did not pass by it because the trail followed the Raft River which is to the south and east of Almo. There is also a feature named "Twin Sisters" four miles past City of Rocks in a south-southwesterly direction past which the trail goes, but neither account mentions it.

[368]"Owned" means "adopted as one's own" *(OED2,* s.v. "own"), i.e., claimed by its owner.

[369]The crow is *Corvus brachyrhynchos.* Up to 20" in length and omnivorous, the crow is smaller than the other big black bird, the raven, and has a shorter, gently-rounded tail (Collins, 168).

[370]Hawks are of the order Falconiformes which they share with vultures and falcons. They are of the family Accipitridae (which they share with kites and eagles), and of genus Accipiter or Buteo. They have a hooked beak and strong claws; they hunt by day and the females are larger than the males. Just what species was seen cannot be determined (Collins, 58-68).

[371]Bennett, 35, goes from Wednesday the 14th to Thursday the 16th! He appears to be missing the real events of the 15th and to have misnamed the 16th Thursday when it was actually Friday. His next entry is described as being for Saturday the 17th, which is correct.

[372]Not mentioned in either Stewart or Holliday – see below where William Fowler Pritchard says that some call it Goose Creek, which is what it is known as today. It is a tributary of the Snake entering it from the south like the Raft River. This is the only place where they were *very* briefly in Utah (for about a mile); the trail comes with a few hundred yards of the Utah border at several points in the seven to ten miles before it reaches Goose Creek but only actually enters Utah while following the creek. Bennett, 35, does not name this river.

[373]Birch is of the family Betulaceae, genus Betula, a type of deciduous shrub or tree, some of which shed their bark, of which there are nine species in the United States. The type William Fowler Pritchard saw was probably mountain birch, a species found to the west of the Rockies, which has shiny reddish-brown bark and small, ovate leaves (Hylander, 182-84).

[374]The word in the manuscript appears to be 'bas' or 'hos' but neither of these is to be found in either regular or slang dictionaries; it is likely that this is what was intended, even if an unusual slang word was being used.

[375]Bennett, 36, says 300'. Stewart, 136, calls these City of Rocks and Cathedral Rocks; he makes this leg 65 miles to William Fowler Pritchard's 64. Clearly, being

well onto Goose Creek, they are beyond what the *Idaho Official Highway Map* (1985) calls "City of Rocks."

[376]Mussels are of the family Mytilidae and are the commonest bivalve. Bivalves constitute the class Pelecypoda, of which there are more than 12,000 species. They have 2 shells (thus bivalve) with a hinge connected by a ligament with one or two muscles. They are mostly sedentary. Both mussels and oysters (see note for 28 July) are of the order filibranchia (Collins, 586-89).

[377]Thousand Springs Creek; there is also a small town named Thousand Springs on highway 93 north of Wells, Nevada *(Official Highway Map of Nevada* (1983)).

[378]Rye is of the *Secale* genus, of Mediterranean origin, which thrives in dry, sandy soil but which was not cultivated until the Middle Ages although the Greeks and Romans knew it (Hylander, 510).

[379]Bennett, 36, says that their neighbors did this and invited them to share it.

[380]Bennett, 36, names the boy Richard and states that he was 10 years old.

[381]Bennett, 36, only mentions the injury to the boy's head.

[382]Bennett, 36, says that there were many other parties encamped at this location and that they thus had nothing to fear from Indians.

[383]Bennett, 36, does not mention this.

[384]Bennett, 36, says that the ox was shot through the nose.

[385]Bennett, 36, does not mention this.

[386]The following text to paragraph end (except for the final words "Very cold evening") comprises two pages in the original journal which occur in the middle of the entry for 19 August (for the exact point, see the note for that day). The words "We keep a very strict watch. I am on" were crossed out just before "Very cold evening" and this extra text has been substituted for these words in this edition. This seems logical because this text continues the discussion of the lack of supplies among other parties.

[387]Stewart, 137, calls it just "Mary's River" and says that this was a new name, in use because of its appearance on Frémont's map. There is presently a Mary's River which runs from the Idaho border due south, then turns west-southwest and flows into the Humboldt (which also runs due south, but a dozen miles further west) *(Official Highway Map of Nevada* (1983)).

[388]The latter part of the entry for this day is confusing; he seems to have conflated these events with earlier ones. Perhaps this is due to the fact that this text is part of an insertion into the entry for another day and is more of a topical than a chronological entry. He says earlier that they travelled 16 miles this day while here 12 miles is specified. None of the material from "We passed through a canyon five miles long" to the end of this entry is in Bennett for this day.

[389]Bennett, 37, says that he was on guard this night as well and specifies 10 p.m. to 1 a.m. as the hours he was on guard.

[390]Like most of their travels in this region, it is hard to know to what this name refers. A "Hot Creek Spring" is about 5 miles northeast of St. Mary's River on the creek of the same name, while "Bathtub Spring" is about 3 miles in the

same direction, but on Badger Creek. Both are about 23 miles into Nevada from the Idaho border (Nevada Department of Transportation, *Nevada Map Atlas,* 5th ed. (Carson City: Nevada Department of Transportation, 1985?)).

[391]This is the point at which the extra material included in the entry for 18 August was inserted in the original.

[392]Bennett, 37, says at this point that they entered "Canon Creek." The *Nevada Map Atlas* shows a Canyon Creek about 10 miles into Nevada which is a tributary of Salmon Falls Creek.

[393]Bennett, 37, says they travelled 28 miles this day.

[394]Bennett, 37, mentions a nasty canyon and hot springs, saying that the canyon lasted for 5 miles and that they travelled 12 miles in all that day.

[395]Bennett, 37, says 4 horses.

[396]"A short gallop or run at full or great speed" *(Webster's,* s.v. "career"). "To turn this way or that." *(OED2,* s.v. "career").

[397]Bennett, 37, says they crossed the Humboldt River at 15 miles.

Stewart, 136, makes this leg of the journey 95 miles to William Fowler Pritchard's 82 (with one day's mileage not reported).

[398]A small group of Indians in southwest Utah and California. Stewart, 137, says these are Shoshones. See also the note for 8 August on the Shoshonis. This branch of the Shoshonis were nomadic gatherers without horses who lived in family rather than tribal groups and were thus quite different than the earlier groups of Shoshonis who were typical plains Indians *(The Encyclopædia Britannica, Micropædia,* 15th ed., s.v. "Shoshoni").

[399]These were probably red fox *(Vulpes fulva)* skins; foxes are of the family Canidae which is also that to which wolves belong. Red foxes have white-tipped tails and may grow to be 42" long; the swift fox *(Vulpes velox)* is smaller (to 31") and has a black-tipped tail and is yellow-brown and nocturnal, while the gray fox *(Urocyon cinereoargentus)* has a black streak down the middle of its tail, grows to 44" and is also nocturnal. The red fox is the most widely distributed, being found in most of the United States and Canada (Collins, 318).

[400]Bennett, 38, gives a better account of the Indians with whom they traded; he says they stayed for dinner. However, he says that there was an earlier party of 12 Indians which they met which Mr. Fewer had encountered down the road and from whom he had a good deal of trouble extracting himself. Much of the following description is not in Bennett.

[401]Bennett, 38, says 15 miles.

[402]Bennett, 38, does not mention this.

[403]Frogs are of order Salientia which they share with toads; there are thirty-three species in the United States. Frogs have smooth skin while toads have rough skin. The family Ranidae are the ordinary frogs, although there are other families of frogs — just which type of frog was seen cannot be determined. Frogs breed in water although some live in dry areas. They hibernate and eat mostly insects (Collins, 434-50).

[404]Bennett, 38, says they asked for food, not tobacco.

[405]Bennett, 38, says they travelled 20 miles this day.

[406]Bennett, 38, does not mention this.

[407]Bennett, 39, says ten dollars.

[408]Stewart, 137-38, says that most boring, remarkable things went unnoticed and unreported.

[409]Bennett, 39, makes the "start" of the day's travel from when the cattle were fed and he thus calls this day an early start, not a late one as William Fowler Pritchard does. He says they travelled 2 miles before breakfast.

[410]Bennett, 39, notes that the Dexter party passed them by on the other side of the river.

[411]Bennett, 39, does not name this tribe.

The Cherokee were at the time of their first contact with Europeans resident in North Carolina and Tennessee. They adopted white customs to an unusual degree (as William Fowler Pritchard notes below) but this did not spare them from the relocation almost all eastern Indians suffered in the early to mid-19th Century. The exact reading and meaning of "Lewissee" is not known; it occurs above the crossed-out word "Arkansaw" which was the original object of "from." Arkansaw is likely a slightly-off geographical reference: they were relocated to northeastern Oklahoma, near Arkansas, in 1838-39. Perhaps this is a reference to the Arkansas River, which flows through the part of Oklahoma to which the Cherokee were sent *(Encyclopædia Britannica, Micropædia,* 15th ed., s.v. "Cherokee").

[412]I.e., the California gold fields.

[413]Of the *Brassica* genus, a headless variety of European ancestry. Other types include broccoli, Brussels sprouts, cabbage, cauliflower, kohlrabi, and rutabaga (Hylander, 246). For related plants, see next note.

[414]*Lepidium* of the family Cruciferae (mustards). Gray, 701, was the only source to identify this as "tongue grass." There are more of this family than of any other except the grasses – more than 2,000 species, mostly weeds. It is an herbaceous family which includes kale (see previous note), mustard, turnip, radish, and horse radish. Also called Pepper Grass or Pepperwort (Hylander, 241).

[415]See the reference for 1 July for mint in general; it is unknown what sort of mint "babsen" mint is.

[416]Of the Sunflower tribe of the Thistle family (see note for 17 June for this family). The Sunflower itself is of genus *Helianthus,* of which there are 100 American species, 20 of which are found on the prairie and 6 of which are found in the West. It is a big yellow ray flower which has a variety of economic uses (Hylander, 479).

[417]Bennett, 39-40, says that this was a bad campsite and that some of the party visited Pullyblank 15 miles behind them and did not return until late at night.

[418]Bennett, 40, does not mention this.

[419]Bennett, 40, says that it was over 20 miles wide and calls it "salaratus ground" and "ashes." He also mentions encountering some packers and selling them a heifer.

[420]Bennett, 40, says that they *joined* Dexter's party.

[421] Bennett, 39, says this happened Tuesday (see 15 August).

[422] Jonathan Jackson is mentioned on the California Argonauts 1850 card. He came to Philadelphia from England in 1838 and had a silk business in New Harmony. His son, Jonathan Paine Jackson, was born on 20 January 1838 in Mack-lesfield, Cheshire, England and performed with his sister Lizzie in a play in Covington, Kentucky – the same place where William Fowler Pritchard and James Bennett pursued theatrical matters before moving to New Harmony in 1847. The Jonathan Jackson on this trip returned to New Harmony in 1852 as mentioned in William Fowler Pritchard's letter to his wife (which makes it clear that Jackson must have *started* his trip back prior to Christmas 1851), then returned to California with his son in 1857, and died in 1871 (1923 remembrances of his son (who died in 1926), on the Local History Card file card for the son).

On his own card, the Jonathan Jackson on the trip is listed as having died on 4 July 1875 at an age which suggests he was born in 1809 or 1810. He is called the founder of the New Harmony Thespian Society, as having had a silk factory in New Harmony, and as being a weaver in 1867.

[423] William Fowler Pritchard says "More" in both cases here. Bennett reports Moore which is more likely as he has been mentioned in this form of the name previously.

[424] Bennett, 40, says 36 hours.

[425] Of the family Cyprinidae, which also includes carp and goldfish (Collins, 536).

[426] Bennett, 40 (28, not 29 August), does not mention William Fowler Pritchard as the seller of this heifer, its price, or that it belonged to the two missing men.

[427] Gray lists no such plant; Hylander, 347, lists the desert shrub *Thamnosma* of the Rutaceae (Rue) family; and there are a number of references in Uphof, 531, all of which are to members of the Pine family. It is unclear what plant is meant.

[428] William Fowler Pritchard spells this "slew." Similar corrections are made twice below.

[429] Bennett, 40, says 32 miles.

[430] It has proven impossible to identify this plant – there just isn't enough information in William Fowler Pritchard's description.

[431] Bennett, 41, states that an earlier company from Ohio had buried him, that he had been killed earlier, and that they had left a note castigating whoever had left him there unburied.

[432] William Fowler Pritchard spells this either "camelion" or "camelian." Of the family Chamaeleonidae, these are the lizards which change color (*The American Heritage Dictionary*, s.v. "chameleon").

[433] Bennett makes no mention of the events from the burial to this point.

[434] William Fowler Pritchard wrote 1 August by mistake.

[435] Bennett, 41, says this was a party of 5-6 from Pullyblank's company, a break-off from Dexter's party.

[436] William Fowler Pritchard has "More" here.

[437]Bennett, 41, includes a long discussion of beggars and their policy towards them at this point.

[438]Bennett does not mention this.

[439]Buffalo grass is of the family Poaceae (grasses), of the genus Bouteloua or Buchloe. This species is one of the "range grasses" used by ranchers in the West to feed cattle and as soil binders (Hylander, 512-14).

[440]Bennett, 41, says they travelled 12 miles.

[441]Bennett, 41, mentions this man in his entry for 1 September.

[442]Bennett, 41, says 15 miles.

[443]Bennett, 41, says 5 miles.

[444]Bennett, 42, says 20 miles.

[445]Bennett, 42, says that they distributed 75 pounds of beef to the neediest; he also remarks that he caught 16 fish this day. William Fowler Pritchard mentions *selling* a quarter of beef the next day and giving away other provisions.

[446]Bennett, 42, says that 15 of these ducks were killed by Mr. Moore. Stewart, 120, says that ducks were very numerous at the time of the southward migration of water fowl.

[447]Bennett, 42, says that this was a boy.

[448]Bennett, 42, says 18 miles.

[449]Bennett does not mention this meeting.

[450]Bennett, 43, also uses this term on his entry for 11 September...

[451]This way of referring to Bennett makes it clear that the James Bennett on the trip with William Fowler Pritchard is a different Bennett than the James Bennett with whom he came from Covington, Kentucky to New Harmony in 1847 and with whom he participated in many theatrical productions. For further discussion of the various James Bennetts, see the Introduction.

[452]Bennett, 42, says it was 150' up the hill.

[453]Bennett, 42, complains that he saw no reason for this full day's rest. This quote is presumably just William Fowler Pritchard's: it is not listed in John Bartlett, *Familiar Quotations* (Boston: Little, Brown and Company, 1980).

[454]Bennett, 42, calls it "Big Meadow." There is no place of this name on the 1:24,000 quadrangles for the area between today's town of Loveland and the Humboldt Sink. The sink itself is a swamp and there are several other possible areas in the wide valley, but none singled out with this name.

[455]Bennett, 42, says that it is 23 miles distant at this point.

[456]Bennett, 43, combines the 8th and the 9th under the entry for the 8th.

[457]*Ignis fatuus,* or "foolish fire," phosphorescent light at night over swampy ground *(Webster's,* s.v. "will-o'-the-wisp").

[458]Bennett, 43, does not mention their tribe. Stewart, 138, says that they had been in Paiute country since the big bend of the Humboldt. This was a Shoshoni speaking tribe in the southwestern United States, the northern branch of which was known as the Diggers; they fought the whites because of the way the emigrants despoiled their lands *(Encyclopædia Britannica, Micropædia,* 15th ed., s.v. "Paiute"). See the note on the Diggers in the entry for 22 August.

[459]Possibly *Sporobolus airoides,* or fine-top salt grass, which is reported on the plains; several other varieties of 'salt grass' are listed, but they are all limited to the Gulf and Atlantic coastal areas (Gray, 154).

[460]Bennett, 43, says 18 miles.

[461]Bennett, 43, counted 64 oxen and 55 horses and mules in one five mile stretch.

[462]Bennett, 43, says 4 a.m.

[463]Bennett, 43, says in his entry for 11 September that Moore was left with a wagon whose team had given out 6 miles short of the Carson River and that he was sent back for it and arrived by 11 a.m. William Fowler Pritchard mentions this in the entry for 12 September.

[464]Bennett, 43, says 11 o'clock.

Stewart says 395 miles, William Fowler Pritchard 346, plus two unrecorded days and one with questionable mileage for this stage. Their arrival in early to mid-September puts them ahead of the usual schedule (unlike their start, which was late).

[465]Bennett, 43, says that these goods were all packed in from Sacramento. Stewart, 297, says that this place was called "Ragtown."

[466]The raven is *Corvus corax,* the largest of several types of black birds, up to 25" in length (the common crow gets no larger than 20"), with a wedge-shaped tail (Collins, 167-168).

[467]Bennett, 43, says 45 miles. Stewart, 138, says the Truckee route was 40 miles and the hardest stretch yet and that everyone arrived tired; the Carson River route which they took was of similar length and difficulty.

[468]Spelled "Kitt" by W. F. Pritchard. Christopher Carson (1809-1868), trapper, trader, and Indian agent, was with Frémont in California during the Bear Flag Revolt which precipitated the Mexican War; he reached the rank of Colonel fighting on the Union side during the Civil War (*Encyclopædia Britannica, Micropædia,* 15th ed., s.v. "Carson, Christopher").

[469]Stewart, 139, says it was used in 1848 (i.e., one year earlier than reported here) and that the Truckee route to the north of this route was the more used.

[470]This word's meaning is unclear: does he mean "he was taking the ferry when we were there" or "he owned the ferry?"

[471]Bennett, 44, worries about snow and talks about needing to pack to make it in time.

[472]This is one of the few places in Bennett's journal that he mentions what William Fowler Pritchard was doing (p. 44). He does not mention the departure of the three people William mentions here, but remarks that Sweasey again wants to stop, this time for three days, which he feels is not a good idea due to the lateness of the season and danger of snow. At this point, the journal of Bennett becomes sketchy because he, Mills, and Lyon joined a cattle speculator on 18 September to drive his herd of 100 cattle over the mountains and he no longer had time to make a daily journal entry and is reporting from memory at some later date. The party they joined was an eleven man party which went by a wagon-negotiable road and

thus he separated from William Fowler Pritchard for the remainder of the journey. There is no way to say for sure whether the three people William Fowler Pritchard mentions are Bennett, Lyon, and Mills, but it is possible, even given the later date Bennett gives for their departure and the difference in the description of who they went with: "some traders" (William Fowler Pritchard) or a cattle speculator (Bennett).

[473]If the President of the United States is meant (which it presumably is), this must be Zachary Taylor, 12th President, who died in office on 9 July 1850, meaning that it was over two months before the party heard of this event (*Encyclopædia Britannica, Micropædia,* 15th ed., s.v. "Taylor, Zachary").

[474]William Fowler Pritchard has "Wollan mits."

[475]William Fowler Pritchard abbreviates this "do."

[476]Stewart, 139, says 210 miles; William Fowler Pritchard's mileage figures add up to more than 200, including one day of 50 (on foot)! The totals for the whole trip are 1970 for Stewart compared with 1918 for William Fowler Pritchard (plus 9 days with no mileage reported, 1 day for which Bennett's figure is used in the absence of one from William Fowler Pritchard, and 5 days of unsure mileage).

[477]He is still in Nevada, between Carson City and Lake Tahoe, at what is now a state historic monument.

[478]Meaning "to have seen something amazing." This phrase, which is said to have come into wide use among soldiers during the Civil War, is thought to have originated among farmers in reaction to seeing elephants at travelling circuses (Robert Hendrickson, *The Facts on file Encyclopedia of Word and Phrase Origins* (New York: Facts on file Publications, 1987), s.v. "seeing the elephant").

[479]See the note on bears for 2 July.

[480]In many of the sentences which follow through 22 September, "I" in the text represents "we" in the original overwritten by "I," presumably editing done by William Fowler Pritchard. Other pronouns have been changed to make this singular usage totally consistent (William Fowler Pritchard's editing was incomplete). His crossed-out words for 20 September (see the note just below) further complicate the interpretation of this period of the trip. He definitely rejoined his recent packing companions on 22 September.

[481]The panther, also known as the mountain lion, cougar, or puma, is *Felis concolor.* It is the largest and only unspotted cat in the United States. It can reach 7½' in length (plus a 3' tail) and 260 pounds in weight (the female is one third smaller). Its main prey is deer (Collins, 330).

[482]Bennett, 45, also reports going over three "ranges."

[483]The words "they sold one of us a little flour" appear here, crossed out, in the manuscript.

[484]I could find no reference to this term in any dictionary I consulted (see Bibliography).

[485]James Fenimore Cooper (1789-1851) was the first major U. S. novelist; he is best known for his Leatherstocking Tales, of which *The Pioneers* (1823) was the first

published but the fourth in narrative order of the series of five *(Encyclopædia Britannica, Macropædia,* 15th ed., s.v. "Cooper, James Fenimore").

[486]The words "came to a ranch and stayed for supper" appear here, crossed out, in the manuscript.

[487]The Colman referred to here is probably George Colman the Younger (1762-1836), a comic dramatist and writer of operas who wrote sentimental works which were box-office successes. Possibly George Colman the Elder (1732-1794) *(Encyclopædia Britannica, Micropædia,* 15th ed., s.v. "Colman, George, the Elder" and "Colman, George, the Younger").

[488]Between the Middle and South Forks of the American River, about 8 miles north of Coloma, which is on the South Fork.

[489]On the South Fork of the American River, some 32 miles east-northeast of Sacramento.

[490]It took Bennett until 1 October to get to Sacramento (Bennett, 45).

[491]The 27th was a Thursday.

[492]I.e., the type of work Jackson was doing, working on the levee.

[493]The following text to paragraph end was found on pages 1-2 of book one — see the Introduction for further information on the manuscript itself.

[494]This appeared in a separate notebook: "Gregory's Express Pocket Letter Book (1851)." Gregory's Express describes itself in this booklet as a company in San Francisco which sent letters to New Orleans and New York City via the Isthmus of Panama. See the Introduction for a photo of the inside cover and first page of this document.

[495]This is where there are two blank pages in the letter book. It is possible that he intended to fill these pages with some further text but never got around to it because the sentence doesn't quite make sense as written.

[496]I.e., the Hawaiian Islands.

[497]I have been unable to identify this theater further; the Eagle Theater (1849) was the first theater in Sacramento *(Greater Sacramento Map* (Sacramento: Chamber of Commerce, 1947)).

[498]I have been unable to identify this person further.

[499]This town is some 50 miles due north of Sacramento on the Feather River, very close to Yuba City.

[500]This is English usage: to fasten, guide, or hold with a guy (i.e., cable); to make an object of ridicule; to exhibit (a person) in effigy like Guy Fawkes *(OED2,* s.v. "guy").

[501]I have been unable to identify this person further. In fact, it is not certain that this is the name of his wife (mentioned again below) — it could be the name of the theater at which his company was engaged or even the name of the company itself.

[502]*OED2,* s.v. "kick up," says "a violent disturbance or row; a great 'to-do', a dance or party" and remarks that this is colloquial usage, originally from the United States — but *The American Heritage Dictionary* does not have this word!

503Almost certainly Charles R. Thorne, Sr. (1814-93), actor and theatrical manager, who went to California in 1850 with his wife and two sons. He managed the American Theater in San Francisco and, later, several theaters in Sacramento *(The National Cyclopedia of American Biography* (New York: James T. White & Co., 1909), 10:401). William Fowler Pritchard spells his name "Thorn."

504I have been unable to identify this person further.

505I have been unable to identify this theater further.

506This town is some 55 miles north-northwest of Sacramento.

507This was Sacramento's finest hotel, built in 1852 at 1018 2nd Street *(Greater Sacramento Map* (Sacramento: Chamber of Commerce, 1947)). The map just calls it the Orleans Hotel. The date given for its construction suggests that this letter may have been written in early 1852 and that William Fowler Pritchard's statement that he would like to have been home by Christmas came true – i.e., that he left in August 1852, later in the year this hotel was built. There is no information on whether he went to the Sandwich Islands or not; if the letter was written in 1852, he hardly had time to go and return before August. Had he gone, one suspects he would have had comparisons to make between what he saw on his way home and what he had seen earlier on his way to and in the Hawaiian Islands, but no such comments appear.

508Or McGraw or Crane. I have been unable to identify this person further.

509Or Hains. I have been unable to identify this person further.

510I have been unable to identify this person further.

511I have been unable to identify this person further.

512I have been unable to identify this person further.

513Or Barrybriss. I have been unable to identify this person further.

514An Albert Fisher is mentioned on the California Argonauts 1850 card. He also has a regular Local History Card file card on which it is stated that he was born in 1818, was a blacksmith in New Harmony in 1844, and died on 2 February 1871.

515A number of Doyles are mentioned on a single card in the Local History Card file. An Alexander Doyle (born in Granard County, Ireland, in 1819, died 24 July 1877 in Smartville, California), had a tin shop in New Harmony from 1843-1848 and married a daughter of M. T. Carnahan. There appears to be another Doyle who also married a daughter of M. T. Carnahan, whose widow, Mrs. E. J. Doyle, died in Marysville, California, in 1885 or 1886. Although brothers marrying sisters was not unusual (two of William Fowler Pritchard's sons married Metcalf sisters), it is at least possible that "E. J." are the wife's initials and her husband was the above-mentioned Alexander Doyle. finally, there is a Charles H. (or M. or N.) Doyle, born in 1850 in New Harmony, died 1878 in Marysville, California, presumably the son of one of these Doyles, mostly likely the son of "Mrs. E. J. Doyle" given the coincidences of their places of death.

516This would be a new Mrs. Bolton (assuming Samuel and not his son William is meant as her husband); the Mrs. Samuel Bolton on the trip died during the trip on 21 July 1850.

[517]See the note for 10 April 1850 for a possible reference to her husband on the trip and a discussion of this name.

[518]Dr. Robert Robson (1801-1878), physician, had a drug store in New Harmony in 1825 (Local History Card file).

[519]Or Read or Real. A Dr. Daniel W. Neal is mentioned in the California Argonauts 1850 card, but the first letter of this name does not look like "N." Given the marriage of a Beal to one of the next-mentioned persons, perhaps this name is Beal, but the first letter does not look like a "B" either.

[520]William Fowler Pritchard spells their name "Leichtenberger." They are listed on the California Argonauts 1850 card. Adam married Caroline Beal, the daughter of John Beal, one of the people definitely with William Fowler Pritchard's party. See note for John Beal (30 June).

[521]There is a William Daniel in the Local History Card file who was a tailor in New Harmony in 1848.

[522]The word looks like "hipe" but if a known word is to be substituted, it is most likely that the "h" is a "tr" with an uncrossed "t." "Pipe" makes sense but just doesn't match what's written.

[523]William Fowler Pritchard spells this name Felsh, but there are a number of references to this family in the Local History Card file, all spelled Felch. The card for Lewis Cass Felch (1851-1921) says that he had a brother named Oscar.

[524]I have been unable to identify this person further.

[525]Son of Samuel and Emma Bolton (see their note for 20 May). His own card in the Local History Card file says that he appeared in plays as a boy in New Harmony and died in California.

[526]Or Hall. John Hale was the son of Lyman and Mary Rogers Hale (who died in 1867); he was a resident of New Harmony most of his life but died of yellow fever in New Orleans (Local History Card file).

[527]The Local History Card file contains the text of the divorce complaint Dr. Cook made against his wife in 1852. He claims he took her to California in the Spring of 1850 because of her "lewdness" and "unchastity," hoping for a change which did not occur. He abandoned her in October 1851 in Nevada City after she took up with a Mr. Roberson who had a wife and children in Missouri. Divorce was granted in September 1852. Another card says that she went to California with a William Robinson, but this appears to be a confusion with later events given the record of the divorce. Mrs. Cook was born Ann Leonora Travers, daughter of James Travers and Elizabeth Patten who were in New Harmony from the 1830's. The date of the trip makes it very likely they were on the train with Bennett and William Fowler Pritchard, although neither mentions this scandalous woman. Perhaps the reference here to her as "lady" suggests that her misbehavior was not well-known at this time. If the date of her abandonment is correct, this letter must have been written in 1851 because they are still mentioned together.

[528]Or Prattorn, or Matton, or Mattorn. The Local History Card file lists a Henry Pratton as being a member of the 1839 Owen geological survey.

529Mt. Vernon is the county seat of Posey County, Indiana, on the Ohio River some 15 miles due south of New Harmony, the starting point of the journey.

There was a John Cunnington (1788-1849) whose daughter called herself Charlotte Conyngton (1819-1863). Charlotte was the mother of Charlotte Emily Metcalf (1853-1930), who married William Fowler Pritchard's son William Shakespeare Pritchard. Because this is an unusual name *and* is geographically related to the Pritchard home area, it is quite likely that the doctor mentioned is some relation to this family.

530Almost certainly William Augustus Twigg (1794-1877). He came to Philadelphia in 1817 and to New Harmony in 1825. His son Charles Augustus is mentioned on the California Argonauts 1850 card. He also had three other sons and a daughter: Eliza Jane Twigg, William Albert Twigg, Alexander Gilbank Twigg, and Robert Dajalma (or Dejalma or Dojalma) Twigg. A Miss Virginia Twigg is listed as the source of this information on the card.

531Looks most like "Vettletran." The Local History Card file lists a Nelson G. Nettleton (1810-1874) as a New Harmony merchant.

532O. D. Chaffee (or Chaffy), a blacksmith, who died in California on 1 January 1858 (Local History Card file).

533Possibly Grillett or Cyullett; but the second letter has a definite looped descender on it, which might be a flourish to the descender of an initial capital "G." The Local History Card file lists a John Gullett who was a member of the Workingmen's Institute and its librarian in 1847 and says that he went to California in 1850, returned, and then took his son Robert (born 1843) with him (by sea this time) to California in 1857, whence he returned in 1863.

534I have been unable to identify this person further.

5351820-1868 (Local History Card file).

536W. F. Pritchard's text has "mat" for the first name. The last name could also be "Stokes." There are several Stokers listed in the Local History Card file: James Madison Stoker (1851-?), listed on the California Argonauts 1850 card, and Richard Stoker who is described as having gone to California in the gold rush and to have been the model for the character "Dick Baker" in Mark Twain's reminiscence *Roughing It* (1872).

537Of gold; it was common in gold fields to pay with gold dust instead of currency.

538W. F. Pritchard's abbreviation is "Cxmas."

539I have been unable to identify this place further (or perhaps it's a literal reference to lions, or a phrase?)

540Junius Brutus Booth (1796-1852), father of the more famous Edwin (1833-1893) and John Wilkes (1838-1865; the assassin of President Lincoln). Edwin eventually became the most famous actor of the three but at the time William Fowler Pritchard was in San Francisco he had been acting for only three years and not regularly; clearly the reference to "the old man" proves that Junius was meant. Both Junius and Edwin were in San Francisco in 1852, where Junius died, probably very shortly after William Fowler Pritchard saw him. Edwin did not become the leading American actor until ca. 1860 *(Encyclopædia Britannica, Micropædia,* 15th ed., s.v. "Booth, Edwin," "Booth, John Wilkes," and "Booth, Junius Brutus," and *Macropædia,* s.v. "Booth, Edwin").

[541]Almost certainly Joseph Proctor (1816-97), actor. He made a tour of the United States in 1851-59 which included California *(The National Cyclopedia of American Biography,* 15:47). William Fowler Pritchard spells his name "Prockter."

[542]Another reading of the last word in the manuscript might be Mane or Wane. This is almost certainly John Thomas Haines' *The Wizard of the Wave; or, the Ship of the Avenger,* a three act play presented for the first time in London in 1840 and published in the same year. Haines wrote dozens of plays *(Dictionary Catalog of the Research Libraries of the New York Public Library 1911-1971* (New York: The New York Public Library, 1979), v. 784, and *The National Union Catalog Pre-1956 Imprints* (London: Mansell, 1972), vol. 226). Fayette Robinson also wrote a "romance" with this title but as the information in these very comprehensive catalogs indicate that this was published only in 1853 and what William Fowler Pritchard attended was a play, the first reference is almost certainly correct.

[543]This might be Alcatraz Island or Treasure Island or the (unnamed) island over which the Bay Bridge now passes; the dock from which he left is clearly on the inside part of the San Francisco peninsula because there are no islands on the Pacific side and the next island he mentions lies off the *eastern* tip of the peninsula immediately north of San Francisco. This and other geographical notes relating to the trip home are abased on the *National Geographic Atlas of the World.*

[544]See the previous note.

[545]William Fowler Pritchard calls these the "Frealonies." They are a group of small islands about 30 miles due west of San Francisco.

[546]The albatross is of the family Diomodeidae whose largest members may attain a wingspan of 12'. They are found mainly in the southern hemisphere and north Pacific. There are 14 species in one or three genera (this is a matter of dispute among experts). They are the most marine of all birds and have nostrils enclosed in tubes. There are several possible species William Fowler Pritchard may have seen (Ralph S. Palmer, ed., *Handbook of North American Birds,* 1 (New Haven and London: Yale University Press, 1962), 116-36).

[547]Porpoises are of family Delphinidae which they share with dolphins and blackfish. They have teeth in both jaws and a dorsal fin near the middle of their backs. They are mammals of the order Cetacea which they share with whales (Collins, 310-13).

[548]Whales are of the order Cetacea which are marine mammals, not fish. They have a blowhole high in the head and horizontal tails. There are many varieties, so which type was seen here cannot be determined (Collins, 305-16).

[549]The flying fish is of the family Exocoetidae and has enlarged pectoral or pelvic fins which are capable of giving them the ability to fly low over the water for brief periods *(The American Heritage Dictionary,* s.v. "flying fish").

[550]General Winfield Scott (1786-1866), Whig candidate for President in 1852. He lost to Franklin Pierce. He served as a general in the War of 1812, the Mexican War, and the Civil War (*Encyclopædia Britannica, Micropædia*, 15th ed., s.v. "Scott, Winfield").

[551]The port side (i.e., left when facing the bow of the boat).

[552]This is the southernmost tip of Baja California, Mexico.

[553]W. F. Pritchard's text has "Land oh!"

[554]A sailing ship with two masts with square-rigged sails, two or more triangular headsails, and a quadrilateral sail behind the rearmost mast (*The American Heritage Dictionary*, s.v. "brig" and "spanker").

[555]This may have been Teotepec, a mountain 11,647' high north of Acapulco.

[556]This is actually a gulf in southernmost Mexico.

[557]These were probably the volcanoes in Guatemala.

[558]In Pacific salt water, a fish of the genus *Globicephala* (pilot whale); the *Tautoga onitis* (tautog) is an Atlantic black fish (*The American Heritage Dictionary*, s.v. "black fish," "pilot whale," and "tautog"). They are big marine mammals of the same order as Dolphins and Porpoises (Delphinidae), are specifically *Globicephala melaena*, and may reach 28' in length (Collins, 313).

[559]Similar to a brig (see note for 21 August 1852), but lacking a square sail at the lowest level of the rearmost mast (*The American Heritage Dictionary*, s.v. "brigantine").

[560]A small town in southernmost Nicaragua, opposite Lake Nicaragua at its closest approach to the Pacific (about 12 miles).

[561]An old or worthless horse (*The American Heritage Dictionary*, s.v. "rip"); W. F. Pritchard spells it "ripp."

[562]At some point after William Fowler Pritchard had this rough trip through the area, Cornelius Vanderbilt developed the Vanderbilt Route whose mule stage was replaced by a stagecoach leg (*Encyclopædia Britannica, Macropædia*, 15th ed., s.v. "Nicaragua, Lake").

[563]What town meant is unclear; today, Rivas and San Jorge lie on Lake Nicaragua north of San Juan del Sur, while Peñas Blancas lies to San Juan's south.

[564]Hornets are of family Vespidae (vespoid or paper wasps), of sub-family Vespinae which they share with yellow jackets. They are largely black with small yellowish-white markings, are social insects which build paper-like nests (hence the name of their family) in which only the queen overwinters (Borror, 348).

[565]Lake Nicaragua is the largest freshwater lake in Central America, some 3190 square miles in surface area, 100 miles long, and averaging 36 miles wide (*Encyclopædia Britannica, Macropædia*, 15th ed., s.v. "Nicaragua, Lake").

[566]The American alligator is *Alligator mississipiensis* which differs from the crocodile in having a broader, shorter snout. The other type of alligator is found in China (*The American Heritage Dictionary*, s.v. "alligator").

[567]Sharks are of the class Chondrichthyes which they share with skates and rays. This class of fish differs from other fishes in that their entire skeleton is cartilaginous. Most sharks are marine fish but there are freshwater sharks, and William

Fowler Pritchard must have seen one of these. The sharks are of the order Selachii (Collins, 461-67).

[568]This town is on the southeasternmost tip of the lake, where the San Juan River begins its journey to the Gulf of Mexico.

[569]In full, El Castillo de Las Concepción, about one third of the way to the Gulf of Mexico.

[570]A yawl is a ship's small boat.

[571]The Gulf of Mexico terminus of the San Juan River, at the southernmost tip of Nicaragua.

[572]Parrots are of the family Psittacidae which they share with parakeets (which are just small parrots). They are highly specialized and are different enough from other birds to have their own order. There are 82 genera and 316 species. They have short, stout, and hooked bills, they are stout and not good at distance flying, are colorful, arboreal, most vegetarian, gregarious, sedentary, and tropical (although they have spread even to cold areas of South America). They are the noisiest of birds and are known for their ability to mimic human speech (Sir A. Landsborough Thomson, *A New Dictionary of Birds* (New York: McGraw-Hill Book Company, 1964), s.v. "parrot").

[573]The snipe found in North America is *Gallinago gallinago delicata,* a species which migrates to and from India. It is of the family Scolopacidae which are ground-dwelling, migratory, wading birds. It is of sub-family Scolopacinae which it shares with the woodcock. It has a long, straight bill, short legs and neck, and cryptic plumage. It inhabits open marshland (Thomson, s.v. "Sandpiper").

[574]Possibly "plum" is meant, i.e., a pudding or cake made with raisins *(The American Heritage Dictionary,* s.v. "plum"); the manuscript very definitely has a final "b".

[575]W. F. Pritchard spells it "Missippee."

[576]General Edward Packenham, the British commander in this battle fought during the War of 1812 *(Encyclopædia Britannica, Micropædia,* 15th ed., s.v. "New Orleans, battles of").

[577]Andrew Jackson, 7th President of the United States. He was a lawyer, member of the House of Representatives, Senator, judge, and the Major General of the Tennessee militia in the War of 1812 where he won major battles in 1814 and 1815, the latter being this battle of New Orleans *(Encyclopædia Britannica, Micropædia,* 15th ed., s.v. "Jackson, Andrew").

[578]No hour specified – blank in the manuscript.

[579]It is not currently known where the family lived when in New Orleans; they arrived by boat from Liverpool on 22 June 1842 (see the Introduction).

[580]Meaning "gone to their graves," i.e., died. This usage is reported to date from the 14th Century (Paul Beale, ed., *Eric Patrtidge, A Dictionary of Slang and Unconventional English,* 8th ed. (London: Routledge & Kegan Paul, 1984), s.v. "long home, (one's)"). One type of ancient grave in England is the barrow which is often a long, narrow construction; thus the use of "long" in this phrase makes sense.

[581]John Fowler Pritchard, born 20 October 1850.

BIBLIOGRAPHY

Bartlett, John. *Familiar Quotations: A Collection of Passages, Phrases and Proverbs Traced to Their Sources in Ancient and Modern Literature.* 15th ed. Boston: Little, Brown and Company, 1980.

Beale, Paul, ed. *Eric Patrtidge, A Dictionary of Slang and Unconventional English.* 8th ed. London: Routledge & Kegan Paul, 1984.

Bennett, James. *Overland Journey to California: Journal of James Bennett Whose Party Left New Harmony in 1850 and Crossed the Plains and Mountains until the Golden West was Reached.* New York: Edward Eberstadt, 1906. Originally published in *The New Harmony Times,* 16 March-3 August 1906. Reprinted by Ye Galleon Press, Fairfield, Washington, 1987.

Borror, Donald J., and Richard E. White. *A Field Guide to the Insects of America North of Mexico.* Boston: Houghton Mifflin Company, 1970.

Bridges, William. *The Bronx Zoo Book of Wild Animals.* New York: The New York Zoological Society and Golden Press, 1968.

Chesterfield (Derbyshire, England). Church of St. Mary and All Saints. *Schedule of Graves in Churchyard.* 1933.

The Chicago Manual of Style. 13th ed. Chicago: The University of Chicago Press, 1982.

Collins, Henry Hill Jr. *Complete Field Guide to American Wildlife.* New York: Harper and Row, Publishers, 1959.

Dictionary Catalog of the Research Libraries of the New York Public Library 1911-1971. New York: The New York Public Library, 1979.

Ditmars, Raymond L. *A Field Book of North American Snakes.* New York: Doubleday, Doran & Co., Inc., 1949 {overwritten 1945 on the copy in the New York Public Library Research Library}.

Dohnalek, Mary. Personal experience.

Ellesmere (Shropshire, England). Church of the Blessed Virgin Mary. Graveyard.

Elliott, Mrs. Josephine M. Personal communication.

Encyclopædia Britannica. 15th ed. 30 vols. Chicago: Encyclopædia Britannica, 1982.

"Excerpts from New Harmony Newspapers..." *Newsletter of the Descendants of William Fowler Pritchard* 2 (April 1988): 18.

Fernald, Merritt Lyndon. *Gray's Manual of Botany.* 8th ed. New York: American Book Company, 1950.

Hall, E. Raymond, and Keith R. Kelson. *The Mammals of North America.* 2 vols. New York: The Ronald Press Company, 1959.

Hendrickson, Robert. *The Facts on File Encyclopedia of Word and Phrase Origins.* New York: Facts on File Publications, 1987.

Holliday, J. S. *The World Rushed In: The California Gold Rush Experience.* London: Victor Gollancz Ltd, 1983.

Hylander, Clarence J. *The World of Plant Life.* New York: The Macmillan Company, 1939.

Idaho. Transportation Department. *Idaho Official Highway Map.* 1985.

Illinois. *Illinois Highway Map, 1985-86.* 1985.

International Genealogical Index. 1984.

Kansas. Department of Transportation. *Kansas Official Transportation Map, 1984.* 1984.

Mawdsley's Map of the City of Liverpool and Suburbs, 1884.

Missouri. Highway and Transportation Department. *Missouri Official Highway Map, 1987-88.* 1987.

Morris, William, ed. *The American Heritage Dictionary of the English Language.* Boston: American Heritage Publishing Co., Inc. and Houghton Mifflin Company, 1969.

The National Cyclopedia of American Biography. New York: James T. White & Co., 1909 (vol. 10) and 1914, 1916 (vol. 15).

National Geographic Atlas of the World. 6th ed. Washington, D.C.: The National Geographic Society, 1990.

The National Union Catalog Pre-1956 Imprints. London: Mansell, 1972.

Nebraska. Department of Roads. *Nebraska 1985-1986 Official Highway Map.*

Nevada. Department of Transportation. *Nevada Map Atlas.* 5th ed. Carson City: Nevada Department of Transportation, [1985?]

————. Department of Transportation. *Official Highway Map of Nevada.* 1983.

New Harmony. Workingmen's Institute. Local History Card File.

New Harmony Advertiser. 20 August 1859.

Newsletter of the Descendants of William Fowler Pritchard. March 1988-.

The Order for Morning Prayer, Daily Throughout the Year. ca. 1708 ["Queen Anne Prayer Book," containing handwritten Pritchard family genealogical information].

The Oxford English Dictionary. 2nd ed. 20 vols. Oxford: at the Clarendon Press, 1989.

Palmer, Ralph S., ed. *Handbook of North American Birds.* 5 vols. New Haven and London: Yale University Press, 1962.

Pigot's & Co.'s Directory. 1842.

Post Office Directory of Shropshire. 1856.

Pritchard, Earl H. Personal recollections.

Pritchard. Phil. "More on W. F. Pritchard's House in New Harmony." *Newsletter of the Descendants of William Fowler Pritchard* 8 (May 1993): 18.

————. "The Properties of William Fowler Pritchard in New Harmony, Indiana." *Newsletter of the Descendants of William Fowler Pritchard* 2 (April 1988): 4-7, 18.

————. "William Shakespeare Pritchard." *Newsletter of the Descendants of William Fowler Pritchard* 4 (May 1989): 3-4.

Sacramento. Chamber of Commerce. *Greater Sacramento Map.* 1947.

Stewart, George R. *The California Trail: An Epic with Many Heroes.* Lincoln: University of Nebraska Press, 1962.

Taylor, Anne. *Visions of Harmony: A Study in Nineteenth Century Millenarianism.* Oxford: Clarendon Press, 1987.

Thomson, Sir A. Landsborough. *A New Dictionary of Birds.* New York: McGraw-Hill Book Company, 1964.

U.K. Derbyshire. County Records Office. Registers of Baptisms, Burials, and Marriages for Chesterfield.

————. Shropshire. County Records Office. Registers of Baptisms, Burials, and Marriages for Ellesmere.

————. Shropshire. County Records Office. Registers of Baptisms for Hordley.

U.S. Department of the Interior. Geological Survey. Topographical Maps.

————. National Archives. Ship Arrival Lists. New Orleans, Roll 22 (2 November 1841-31 August 1843).

Uphof, J. C. Th. *Dictionary of Economic Plants.* Lehre: Verlag Von J. Cramer, 1968.

Webster's New International Dictionary of the English Language. Springfield, Mass.: G. & C. Merriam Company, 1925.

————. Unabridged ed. Springfield, Mass.: G. & C. Merriam Company, 1981.

Wyoming. State Highway Department. *Wyoming 1987.* 1987.

Index of Persons Mentioned on the Trip Out

[References from the period reported from Bennett's journal are in square brackets]

Lyon, Ira	[1 April]
	[10 April]
	[15 April]
Mills, John	[10 April]
	[19 April]
	1 July
Mitchell	[15 April]
Moore	[20 May]
	29 August
	1 September
Morrison (& wife)	7 August
	31 August
O'Neal, John	[1 April]
O'Neal, Mitch	21 July
Otzman	30 June
	1 July
	21 July
Pullyblank	[10 May]
	21 July
Spencer	[3 April]
	10 September
Sweasey, Richard	17 August
Sweasey, W.J.	[2 April]
	[10 May]
	21 July
	10 August
	17 September
Sweasey, Mrs. W. J.	[1 April]
Wade	[20 May]
Williams, John	21 July
Wilsey, Bill	25 May ("2 Wilseys")
	21 July

Index of Persons Mentioned in His Letter Home

[Names of persons also mentioned on the trip out are followed by a *]
[Information which does not appear in the text is enclosed in square brackets as this sentence is]

O'Neal, Mitch *
Pratton, Henry
Pullyblank *
Pullyblank, Mrs.
Robson, Bob
Stark, Mr.
Stoker, Matt
Thorn[e, Charles R. Sr.]
Twigg, [William Augustus]
Wilkinson, Mr.
Williams, John *
Williams, Mrs. John
Wilsey, Bill *
Wilsey, Mrs. Bill [Lucretia Stoker] (and small child)

Index of Wild Animals Mentioned on the Trip Out

Excluded: fur, feathers, buckskin, game, varmint, bird (generic), fowl (generic), fish (generic), snake (generic)

antelope	31 May
	2 June (description)
	8 June
	9 June
	12 June
	13 June
	23 July
	24 July
bear	2 July
	16 July
	19 Sept (grizzly)
beaver	2 July (description)
buffalo	29 May (robes)
	31 May
	7 June
	8 June
	12 June (dung)
	13 June (chips)
	14 June
	15 June (dung)
	17 June (skull)
	18 June (hides)
	28 June
	3 July (steaks) (comparison to bison)
	22 July (out of their range now)
	29 July (robe)
buffalo fish	25 May
chameleon	31 Aug
crickets	2 Aug
crows	14 Aug
	12 Sept

deer	8 June (comparison to antelope)
	28 June
	2 July
	21 Sept
duck	29 June (wished for)
	2 July (comparison to beaver)
	6 Aug
	7 Aug
	4 Sept
elk	24 May (horns)
	17 June (skull)
	28 June
	2 July
	12 July
	5 Aug
fox	22 Aug (mountain, skins)
frog	23 Aug (toad)
geese	6 Aug
goat	2 June (comparison to antelope)
gopher	14 June (description)
grasshopper	30 June (dirty red)
hare	3 July
	24 July
hawk	14 Aug
hoop snake	15 July
marmot	14 June
minnow	29 Aug
mosquito	21 June
	1 Aug
	2 Aug
mussel	16 Aug
owl	14 June
oyster	28 July (geology, description)
panther	19 Sept (rumor)
pike (fish)	25 May
prairie dog	11 June
	14 June (description)
rattlesnake	20 May
	14 June

raven	12 Sept
rock bass	2 Aug
sage fowl	24 July
sheep, mountain	5 July
suckers (fish)	2 Aug
toad, horned	30 June
trout	27 July (black, not red spots)
	2 Aug
wolf	1 June
	12 June
	13 June
	22 June (2nd hand)
	25 June
	30 June
	9 July (every night + inferred)
	12 July
	20 July (misc.)
	28 July
	25 Aug
	17 Sept
	19 Sept (rumor)

Index of Wild Animals Mentioned on the Trip Home

Index of Wild Plants Mentioned on the Trip Out

[Excluded: wood, timber, feed, roots, forest, foliage, apples, beans, corn, grass, tobacco, tree (generic)]

Absynthea	see greasewood
Artemisia	see sage
asp (tree)	13 Aug
birch	15 Aug
box elder	5 July
buttercups	12 June
cactus	12 June (prickly pear)
	14 June
	23 June
cedar	11 June
	17 June
	22 June
	28 June
	14 July
chamomile	17 June
cottonwood	28 May
	9 June
	5 July (bitter)
	9 July
cress	27 Aug (tongue grass)
currants	30 June
	1 July
	14 July (red)
	1 Aug (red & yellow)
cypress	4 Aug
flax	13 Aug
gooseberries	30 June
	1 July
	14 July
	24 July (pies)
grape, wild	4 June
greasewood	12 July (Absynthea)
	15 Aug
kale	27 Aug (cabbage)

lavender	17 June
mint	1 July
	27 Aug (babsen)
mushrooms	5 June
	11 June
oak	20 Sept
	24 Sept
	27 Sept
oats, silver	13 Aug
pea, wild	5 June
pine	22 June
	28 June
	15 July
	19 July
	4 Aug
	18 Sept
	20 Sept
prickly pear	see cactus
resin plant	30 Aug (turpentine)
rye, wild	17 Aug
	19 Aug
sage	17 June (Artemisia)
	29 June (wild)
	30 June (wild)
	1 July (wild)
	10 July (wild)
	11 July (wild)
	12 July (wild & not wild + description)
	22 July
	26 July (wild)
	16 Aug
sunflower	27 Aug
thyme	1 July
turpentine plant	see resin plant
vetch	5 June

willow	9 June (twigs & green)
	25 June (brush)
	2 Aug
	15 Aug
	29 Aug
	6 Sept
wormwood	17 June
	18 Aug
unknown	31 Aug ("curious plants full of water")

GENERAL INDEX

COLOPHON

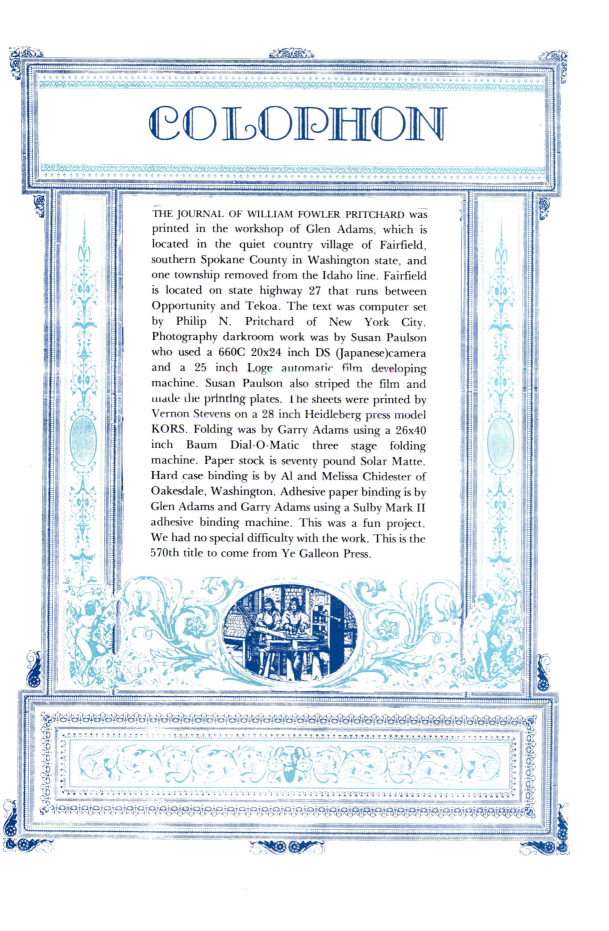

THE JOURNAL OF WILLIAM FOWLER PRITCHARD was printed in the workshop of Glen Adams, which is located in the quiet country village of Fairfield, southern Spokane County in Washington state, and one township removed from the Idaho line. Fairfield is located on state highway 27 that runs between Opportunity and Tekoa. The text was computer set by Philip N. Pritchard of New York City. Photography darkroom work was by Susan Paulson who used a 660C 20x24 inch DS (Japanese) camera and a 25 inch Loge automatic film developing machine. Susan Paulson also striped the film and made the printing plates. The sheets were printed by Vernon Stevens on a 28 inch Heidleberg press model KORS. Folding was by Garry Adams using a 26x40 inch Baum Dial-O-Matic three stage folding machine. Paper stock is seventy pound Solar Matte. Hard case binding is by Al and Melissa Chidester of Oakesdale, Washington. Adhesive paper binding is by Glen Adams and Garry Adams using a Sulby Mark II adhesive binding machine. This was a fun project. We had no special difficulty with the work. This is the 570th title to come from Ye Galleon Press.

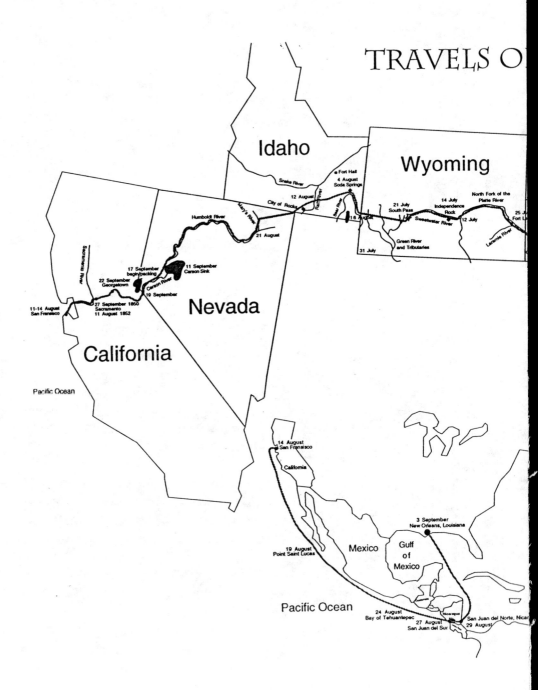

TRAVELS O

Idaho

Wyoming

Snake River · Fort Hall
4 August
Soda Springs

12 August
City of Rocks

Humboldt River

Mary's River

Snake River

Raft River

16 August

21 July
South Pass

14 July
Independence
Rock

Sweetwater River

North Fork of the
Platte River

12 July

25 Ju
Fort L

Laramie River

Green River
and Tributaries

31 July

21 August

Sacramento River

17 September
begin/packing

11 September
Carson Sink

22 September
Georgetown

Carson River

19 September

Nevada

27 September 1850
Sacramento
11 August 1852

11-14 August
San Fransico

California

Pacific Ocean

14 August
San Fransico

California

3 September
New Orleans, Louisiana

19 August
Point Saint Lucas

Mexico

Gulf
of
Mexico

Pacific Ocean

24 August
Bay of Tehuantepec

Nicaragua

27 August
San Juan del Sur

San Juan del Norte, Nicar
29 August

ILLIAM FOWLER PRITCHARD,
1850-52

Nebraska

Fork to North Fork of the Platte

Illinois

Indiana

Kansas

New Harmony

Kentucky

Missouri

Tennessee

Arkansas

Mississippi River

Mississippi

Louisiana

4 September
New Orleans